Finite Automata and Regular Expressions: Problems and Solutions

Stefan Hollos and J. Richard Hollos

Finite Automata and Regular Expressions: Problems and Solutions
by Stefan Hollos and J. Richard Hollos
Paper ISBN 978-1-887187-16-9
Ebook ISBN 978-1-887187-17-6

Abrazol Publishing

an imprint of Exstrom Laboratories LLC
662 Nelson Park Drive, Longmont, CO 80503-7674 U.S.A.

About the Cover

Cover created with the help of POV-Ray and Inkscape. We thank the creators and maintainers of this excellent software.

Contents

Preface

This is a book about solving problems related to automata and regular expressions. There is very little theory here. We cover a few interesting classes of problems for finite state automata and then show some examples of infinite state automata and recursive regular expressions. The final problem in the book involves constructing a recursive regular expression for matching regular expressions.

You will probably get the most out of this book if you already know something about automata and regular expressions. Since you were curious enough to look at the contents of this book, you most likely at least know what they are. The introduction does provide some background information on automata, regular expressions, and generating functions. Out of these three topics the last is the one that most people are probably least familiar with. The generating function associated with an automaton and regular expression will tell you how many strings of a given length are accepted by the automaton.

The inclusion of generating functions is one of the unique features of this book. Few computer science books cover the topic of generating functions for automata and there are only a handful of combinatorics books that mention it. This is unfortunate since we believe the connection between computer science and combinatorics, that is opened up by these generating functions, can enrich both subjects and lead to new methods and applications. The generating functions for infinite state automata are especially interesting and ripe for investigation. The last problem in the book for example, shows how to derive the generating function for the number of syntactically correct regular expressions of a given length.

After the introduction comes a section on using automata to solve divisibility problems. Given a representation for integers in binary, octal, decimal, hexadecimal, etc. it is possible to construct an automaton that will test if an integer is divisible by some other integer. It is also possible to test for divisibility by one integer but not by another and so on.

Next comes a section on pattern matching in strings. This includes general problems on finding strings where runs of some symbols are required or restricted or particular numbers of symbols must occur. There is a de-

3

scription of how to construct automata for efficiently matching a particular pattern in a string. This is essentially the Knuth-Morris-Pratt string searching algorithm, although we do not go into the details of the algorithm and restrict ourselves to constructing automata, regular expressions and generating functions. Problems on three and four bit patterns in binary strings are worked out along with some patterns in ternary and quaternary strings. Some of the more challenging problems in this section are the ones involving patterns in circular strings. These are strings with symbols arranged in circular order with no beginning or end.

The final section is called miscellaneous problems and it contains some of the most challenging problems in the book. Some of the problems are of a more combinatorial nature. There is one related to the coupon collector problem in combinatorics and probability. There are a few problems related to random walks and the gamblers ruin problem. The problems are generalized in a way that provides an easy introduction to the subject of infinite state automata and recursive regular expressions. For example the problem of correctly nested parentheses is first solved in the case where the nesting depth is restricted to 5. We then show how to remove the restriction and how this changes the regular expression into a recursive form and how the generating function can also be solved for recursively. Several infinite forms of the gamblers ruin problem are also solved. The final problem deals with finding a recursive regular expression for matching regular expressions over a binary alphabet. A list of all the syntactically correct regular expressions of length 1, 2, 3, and 4 are given in the solution. The recursive regular expressions in this section are also a nice way of introducing the topic of context free grammars. The context free grammars for some of the recursive regular expressions in these problems are also given.

Introduction

Automata

Why are automata important? One reason is that when you need a regular expression, sometimes it's easier to create the automaton, then get the regular expression from that. So being able to make an automaton for what you want is a precious skill. Secondly, some programs that take regular expressions as input, like grep and flex, will convert that regular expression into an automaton, then use that with a very fast algorithm to perform the search. Thirdly, automata are an elementary computational model, as we explain below.

An automaton is a very simple little computer. It has a set of states and it transitions from one state to another according to the sequence of inputs it reads. The inputs are usually represented as a string of symbols or characters. One of the states is the start state and one or more of the states are end states. The start state may also be an end state. The computation it performs is a yes or no answer to the question: does the input sequence move the automaton from the start state to one of the end states? The answer will of course depend on how the states are connected. To see clearly how all this works requires some formal notation and definitions.

Let Q represent the set of states of an automaton and let $q_i \in Q$ be an element of Q. At the end of the book we will introduce some problems that require automata with a countably infinite number of states but most of the automata in this book will have a finite number of states. Let $q_0 \in Q$ be the start state. The automaton always starts in this state. Let $E \subseteq Q$ be the set of end states. If the automaton is in a state $q_i \in E$ at the end of the input then the input is said to be accepted. The end states are also sometimes called accept states or final states. Let Σ be a finite set of symbols or characters called the alphabet. The input sequence will consist of a string of characters taken from Σ. Let $\delta(q_i, a) = q_j$ be the automaton transition function. The function says that if the automaton is in state q_i and the next input symbol is $a \in \Sigma$ then the automaton will move to state q_j. If the automaton is in state q_i and it reads a symbol for which no transition is defined then it stops and the input is not accepted.

Human beings like pictures and for that reason automata are usually represented as directed graphs. Flip through the following pages and you will see lots of pictures of automata. The nodes in the graph represent states and the directed edges represent transitions from one state to another. Each edge is labeled with an alphabet symbol. If $\delta(q_i, a) = q_j$ then there will be an arrow from node q_i to node q_j labeled a. You will sometimes see start nodes designated by a double circle. In this book we will simply state what the start and end nodes are in those cases where it is not obvious.

You will also sometimes see edges labeled with the symbol ϵ. This is the symbol for an empty string and it means that the transition is possible with no input. If an automaton has ϵ transitions or if it has a state with two different transitions for the same input symbol then it is called nondeterministic. Otherwise it is called deterministic. The abbreviations NFA and DFA are used for nondeterministic and deterministic finite automaton. Every NFA can be turned into an equivalent DFA with a larger number of states. For the problems in this book the distinction between an NFA and a DFA will not be important. More information about their differences and how they are related can be found in references at the end of the book.

One of the things we will want to do in the following problems is to characterize the family of strings accepted by an automaton. One way to characterize them is by what is called a regular expression (discussed more fully in the next section). To find the regular expression we will need to simplify the automaton by removing some of its states. As an example look at the three state automaton in the figure below.

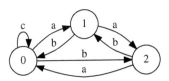

What we want to do is remove state 0, leaving only states 1 and 2. The transitions in the reduced automaton should be such that state 0 is still effectively present. The reduced automaton is shown in the figure below.

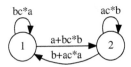

You can see that the transitions are no longer labeled by single symbols. The transition from state 1 to 2 is now labeled $a + bc^*b$ where the $+$ operator is an OR operator and the $*$ superscript on the c indicates any number of c's, $c^* = \epsilon + c + cc + ccc + \cdots$. The label means the transition can occur on a single a or on a b followed by any number of c's, followed by another b. It is the bc^*b part of the transition that takes into account the effect of the removed 0 state. You can see from the original automaton that it is possible to go from state 1 to state 2 by first going to state 0 on a b input, staying at state 0 for any number of c inputs and then going to state 2 on another b input. When state 0 is removed this possibility has to be added to the transition from state 1 to 2. In the original automaton you can return to state 1 by going through state 0 on an input of bc^*a so this is taken into account by the loop back transition on state 1 in the reduced automaton.

You could continue in this example by eliminating state 2. This would leave only state 1 with a loop back transition labeled $bc^*a+(a+bc^*b)(ac^*b)^*(b+ac^*a)$. If state 1 is both the start and end state for this automaton then the strings it accepts are the ones that can be generated by going around this loop transition any number of times. The strings are then characterized by the expression $(bc^*a + (a + bc^*b)(ac^*b)^*(b + ac^*a))^*$ which is called the regular expression for the automaton.

Regular Expressions

Regular expressions were introduced in the previous section. They show up as labels for transitions when states are removed from an automaton. In this section we are just going to give a slightly more formal definition of regular expressions that may be useful for better understanding some of the problems in this book. For a more detailed discussion see one of the references at the end.

A regular expression is used to define a set of strings constructed from a finite alphabet of letters or symbols. The set of strings so defined is called a regular language. The best way to understand what this means is by example. Let the alphabet consist of four letters $A = \{a, b, c, d\}$. One simple language is all the two letter strings that begin with the letter a. The regular expression defining the language is: $R = aa+ab+ac+ad$. The $+$ operator in the expression is analogous to a logic *or* operator or a set theory union operator. The expression can be simplified a bit by writing it as $R = a(a + b + c + d)$ which represents the concatenation of a with any of the possible one letter strings.

This example shows the two basic operations used to construct regular expressions: union and concatenation. If R and S are regular expressions then $R+S$ and RS are also regular expressions. The simplest regular expression is just a single letter, $R = a$ or $R = \epsilon$ where ϵ

is an empty string i.e. a string with no letters. Multiple concatenations can be written in a manner analogous to exponentiation, $ababab = (ab)^3$ or if $R = ab$ then $RRR = R^3$. Note that $(ab)^3 \neq a^3b^3 = aaabbb$.

Another operator used in regular expressions is called the Kleene star. It is defined as follows: $R^* = \epsilon + R + R^2 + R^3 + \cdots$. A starred regular expression can appear zero or more times concatenated with itself. Using the Kleene star, the regular expression for all possible strings in the alphabet $A = \{a, b, c, d\}$ is simply $R = (a + b + c + d)^*$.

For our purposes this sums up how regular expressions are defined and what they mean. There is some additional shorthand notation that is sometimes used to simplify regular expressions. For example, the regular expression $0 + 1 + 2 + 3 + 4 + 5 + 6 + 7 + 8 + 9$ can be abbreviated as $[0 - 9]$ and the letters a through z can be abbreviated as $[a - z]$. Another common abbreviation is $\epsilon + a = a?$. A simple identity that is sometimes useful is $\epsilon + aa^* = a^*$. For more identities, theorems, and other goodies involving regular expressions see one of the references.

Keep in mind that the language accepted by any finite automaton can be described by a regular expression. The regular expression need not be unique however. It is possible to have two different regular expressions that describe the same language. When eliminating states in an automaton, the order in which they are eliminated can sometimes lead to different regular expressions that are equally valid. This is analogous to the fact that you can have two polynomials that are equivalent, with one being a simplified form of another where redundant terms have been canceled. There is an algebra for regular expressions that can, in principle, be used to simplify them but it's generally not easy to use and not worth the trouble.

Some of the problems at the end of the book will involve automata having an infinite number of states. The languages accepted by these automata cannot be described by regular expressions. They can however be described by a recursive regular expression. An example of a recursive regular expression for strings over an alphabet $A = \{a, b\}$ is $R = (aRb)^*$. The regular expression is defined in terms of itself which allows it to be infinitely long. This is useful for finding the generating function of an infinite automaton. Generating functions are discussed in the next section.

Generating Functions

The language accepted by an automaton can have strings of many different lengths. It is often useful to know how many strings of a given length are in a language. You can do this by constructing what is called a generating function for the language. We will usually use the notation $G(z)$ for the generating function. For a regular language, i.e. a set of strings accepted by a finite automaton, $G(z)$ will always end up being either a polynomial or a rational function of two polynomials which can be expanded as a power series. Some of the problems at the end of the book will involve automata with an infinite number of states in which case the language is no longer regular and the generating function may involve square roots of polynomials. In any case the coefficient of z^n in the power series expansion of $G(z)$ will equal the number of strings in the language that have length n

$G(z)$ can be found directly from the regular expression by replacing each alphabet letter by z and replacing each starred expression, R^* by $(1 - R(z))^{-1}$ where $R(z)$ is gotten by replacing each alphabet letter in R by z. The $+$ operators are then treated as addition operators and concatenation is treated as multiplication. For example the generating function for the language defined by $R = (a + b)(c + d)^*$ becomes

$$G(z) = (z + z)(1 - z - z)^{-1} = \frac{2z}{1 - 2z}$$

If you expand this as a power series you get

$$G(z) = 2z + 4z^2 + 8z^3 + 16z^4 + 32z^5 + 64z^6 + 128z^7 + 256z^8 + \cdots$$

The coefficient of z^n is 2^n, meaning there are 2^n strings of length n in this language.

The generating function for a recursive regular expression can be found in basically the same way. In the previous section we gave $R = (aRb)^*$ as an example of a recursive regular expression. If $G(z)$ is the generating function for R then the expression translates into

$$G(z) = \frac{1}{1 - z^2 G(z)}$$

Solving this equation for $G(z)$ gives:

$$G(z) = \frac{1 - \sqrt{1 - 4z^2}}{2z^2}$$

This expression can then be expanded as a power series with the coefficient of z^n being the number of strings of length n in the language.

Divisibility

Automata can be used to test the divisibility properties of integers. For example, you can construct an automaton that accepts all integers divisible by 3. You can also test for divisibility by one number but not another, divisibility by 3 but not by 2 for example. The structure of the automaton will depend on how the numbers are represented. Testing numbers in decimal form is different than testing them in binary form. Most of the following problems will use numbers in binary form since the automaton tends to be simpler with only two possible transitions out of each state.

The bits or digits of the number are scanned left to right the way they are normally written. This means the most significant bit (MSB) is scanned first and the least significant bit (LSB) is scanned last. Let n be the value of the number at some point in the scanning process and let its remainder upon division by d be r. This means n can be written as $n = md + r$ for some integer m. The remainder can have values $r = \{0, 1, \ldots, d - 1\}$. These remainder values will be used for the states of the automaton.

As more digits of a number are scanned, its value will change and so will its remainder upon division by d. After scanning a new digit, the new remainder will depend only on the old remainder and the value of the new digit. This is what makes it possible to use an automaton for this problem. An automaton can remember only one thing, its current state, which in this case is the current value of the remainder. The automaton is constructed so that it has transitions to a new remainder state depending on the value of the new digit.

In terms of binary numbers let b be the value of the next scanned bit, then the new value of the binary number will be

$$2n + b = 2(md + r) + b$$
$$= 2md + 2r + b$$

So the new remainder will be $r' = (2r + b) \bmod d$ and there will be an automaton transition from state r to r' on input b. There are only two transitions out of each state since there are only two possible values, $b = 0$ or $b = 1$.

In the following problems the start state is always the one labeled 0 and this is also the only accepting state. Regular expressions for the automata are also derived.

Divisibility Problems

Problem 1. Create an automaton for recognizing binary numbers divisible by 2. The bits are to be read from left to right. Find the regular expression.

Answer. Let b be the value of the next bit, then the state will switch from r to r' according to the equation $r' = (2r + b) \bmod 2$. The table below shows the new state r' as a function of r and b. The table is used to construct the automaton shown in the following figure.

r	b 0	1
0	0	1
1	0	1

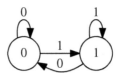

State 0 is both the start and accepting state. Looking at the automaton, it's clear that a number is only accepted if the last bit read is zero. What comes before the last bit can be any binary number. When you divide a binary number by 2, you are essentially shifting all the bits one bit to the right. The least significant bit then becomes the first bit to the right of the decimal. This means the LSB has to be zero for the whole number to be divisible by 2. One could argue then that the regular expression must be

$$R = \epsilon + (0 + 1)^*0$$

and this does indeed work. The regular expression can also be found from the automaton. When a 1 is read, the automaton leaves state 0 and returns after any number of additional 1's followed by a 0. The regular expression for such a string is 11*0. Any number of repeats of this string will put us back in state 0. Also any number of repeats of 0 will keep us in state 0. The regular expression, according to the automaton must then be

$$R = (0 + 11^*0)^*$$

Both this and the above regular expression work and are therefore equivalent.

Problem 2. Create an automaton for recognizing base 10 numbers divisible by 2. The digits are read left to right. Find the corresponding regular expression.

Answer. Let d be the value of the next digit, then the state will switch from r to r' according to the equation $r' = (10r + d) \bmod 2$. The table below shows how the states switch depending on the value of d. The table is used to construct the automaton shown in the figure below.

	d									
r	0	1	2	3	4	5	6	7	8	9
0	0	1	0	1	0	1	0	1	0	1
1	0	1	0	1	0	1	0	1	0	1

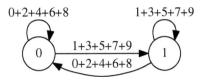

The automaton has the same basic form as the division by 2 in binary, and the regular expression is also similar:

$$R = (A + BB^*A)^*$$

with $A = 0+2+4+6+8$ and $B = 1+3+5+7+9$. Another way to get a regular expression is by noting that the number has to end with a 0, 2, 4, 6, or 8 to be divisible by 2. So the following regular expression will also work.

$$R = \epsilon + [0 - 9]^*(0 + 2 + 4 + 6 + 8)$$

Problem 3. Create an automaton for recognizing binary numbers divisible by 4 and find the regular expression. The bits are read left to right.

Answer. Let b be the value of the next bit, then the state will switch from r to r' according to the equation $r' = (2r + b) \bmod 4$. The table below shows how the states switch depending on the value of b. The table is used to construct the automaton shown in the figure below.

	b	
r	0	1
0	0	1
1	2	3
2	0	1
3	2	3

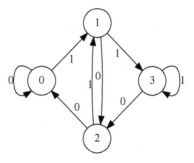

The automaton can go directly from state 1 to 2 by reading a 0. It can also go there through state 3 by reading one or more 1's followed by a 0. This suggests that state 3 can be eliminated by putting a 1 loop on state 1. The simplified automaton is then

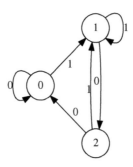

The way to interpret this automaton is that division by 4 is division by 2 twice so the last two bits read must be 0. If a 1 is read we need to read two 0's to get back to the accepting state 0. To get the regular expression from this automaton we eliminate state 2 by connecting state 1 to 0 by the string 00 and adding a loop back to state 1 by the string 01. This gives

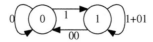

from which it is clear that the regular expression must be

$$R = (0 + 1(1 + 01)^*00)^*$$

An equivalent regular expression is

$$R = \epsilon + 0 + (0 + 1)^*00$$

This comes from the fact as noted above, that to be divisible by 4, the binary number only has to end with two 0's.

Problem 4. Show that the minimum number of states required to test if a binary number is divisible by 2^n is $n + 1$.

Answer. If you divide a binary number by 2^n you shift the bits to the right by n places. The last n bits will end up to the right of the decimal point and must therefore be zero. This means that after reading a 1 we need to read n consecutive 0's to return to the accepting state 0. So we need one state for reading the 1 and n states for reading the n 0's for a total of $n+1$ states. As an example the automaton for testing divisibility by $2^4 = 16$ would look like

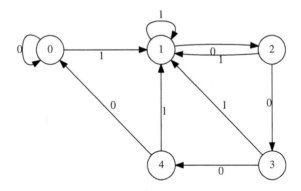

where state 0 is both the start and the accepting
state.

Problem 5. Find the automaton and regular expression for binary numbers divisible by 2 but not by 2^n for $n > 1$.

Answer. For a binary number to be divisible by 2 but not by $4, 8, 16, \ldots$, it must end with 10. The automaton for recognizing numbers that end with 10 is

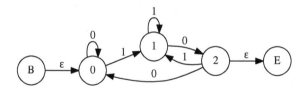

Eliminating state 1 gives

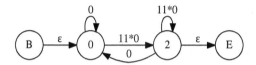

And eliminating state 0 leaves

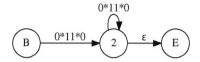

Reading off the regular expression from this automaton gives $R = 0^*11^*0(0^*11^*0)^*$

The 6-bit binary numbers that satisfy this regular expression, along with their base-10 equivalent are:

```
000010 2
000110 6
001010 10
001110 14
010010 18
010110 22
011010 26
011110 30
100010 34
100110 38
101010 42
101110 46
110010 50
110110 54
111010 58
111110 62
```

The numbers divisible by 2^n for $n > 1$ are missing as they should be.

Problem 6. Create an automaton to recognize binary numbers divisible by 3 and find the regular expression. The bits are read from left to right.

Answer. Let b be the value of the next bit, then the state will switch from r to r' according to the equation $r' = (2r + b) \bmod 3$. The table below shows how the states switch depending on the value of b. The table is used to construct the automaton shown in the figure below.

| | | b | |
r	0	1
0	0	1
1	2	0
2	1	2

To get the regular expression from this automaton we eliminate state 2 by adding a loop back to state 1 by the string 01*0. This gives

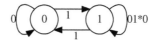

from which it is clear that the regular expression must be

$$R = (0 + 1(01^*0)^*1)^*$$

Problem 7. Create an automaton to recognize base 10 numbers divisible by 3 and find the regular expression. The digits are read from left to right.

Answer. Let d be the value of the next digit, then the state will switch from r to r' according to the equation $r' = (10r + d) \bmod 3$. The table below shows how the states switch depending on the value of d. The table is used to construct the automaton shown in the figure below.

	d									
r	0	1	2	3	4	5	6	7	8	9
0	0	1	2	0	1	2	0	1	2	0
1	1	2	0	1	2	0	1	2	0	1
2	2	0	1	2	0	1	2	0	1	2

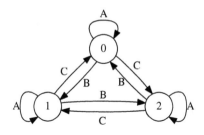

where $A = 0 + 3 + 6 + 9$, $B = 1 + 4 + 7$ and $C = 2 + 5 + 8$. Eliminating state 2 leaves

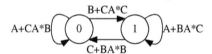

From this you can read off the regular expression:

$$R = (A + CA^*B + (B + CA^*C)(A + BA^*C)^*(C + BA^*B))^*$$

Problem 8. Create an automaton and regular expression to recognize binary numbers divisible by 3 but not by 2.

Answer. Another way to state this problem is that we want to find binary numbers divisible by 3 that do not end with a zero. This can be done with a simple modification of the divide by 3 automaton. We just have to add one more state to ensure that the number does not end with a zero. The automaton is shown below.

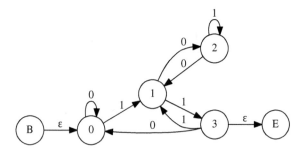

Removing state 2 gives

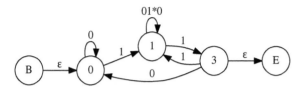

Then removing state 1 gives

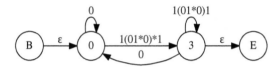

Finally, removing state 0 leaves

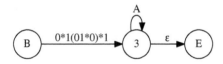

where $A = 1(01^*0)1 + 00^*1(01^*0)^*1$ which simplifies to $A = 0^*1(01^*0)^*1$.

The regular expression can be read off from the last automaton as

$$R = 0^*1(01^*0)^*1(0^*1(01^*0)^*1)^*$$

The 6-bit binary numbers that satisfy this regular expression, along with their base-10 equivalent are:

```
000011 3
001001 9
001111 15
010101 21
011011 27
100001 33
100111 39
101101 45
110011 51
111001 57
111111 63
```

Problem 9. Create an automaton and regular expression to recognize binary numbers not divisible by 3.

Answer. The automaton is basically the divisibility by 3 automaton, but now the end states are 1 and 2 instead of 0. Putting in explicit begin and end states, the automaton is

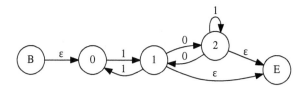

Eliminating state 0 gives

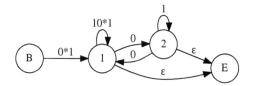

Then eliminating state 2 leaves

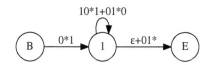

Now we can read off the regular expression as

$$R = 0^*1(10^*1 + 01^*0)^*(\epsilon + 01^*)$$

The 4-bit binary numbers that satisfy this regular expression, along with their base-10 equivalent are:

```
0001 1
0010 2
0100 4
0101 5
0111 7
1000 8
1010 10
1011 11
1101 13
1110 14
```

Problem 10. Create an automaton to recognize binary numbers divisible by 5 and find the regular expression. The bits are read from left to right.

Answer. Let b be the value of the next bit, then the state will switch from r to r' according to the equation $r' = (2r + b) \bmod 5$. The table below shows how the states switch depending on the value of b. The table is used to construct the automaton shown in the figure below.

$$
\begin{array}{c|cc}
 & \multicolumn{2}{c}{b} \\
r & 0 & 1 \\
\hline
0 & 0 & 1 \\
1 & 2 & 3 \\
2 & 4 & 0 \\
3 & 1 & 2 \\
4 & 3 & 4 \\
\end{array}
$$

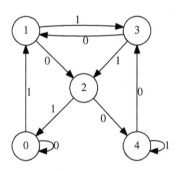

To get the regular expression from this automaton we start by eliminating state 2. This gives

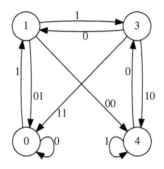

Next we eliminate state 4

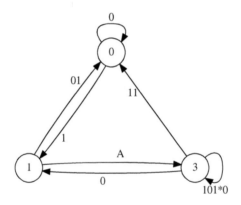

where $A = 1 + 001*0$. Finally, we remove state 3

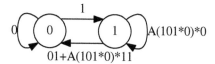

from which it is clear that the regular expression must be

$$R = (0 + 1(A(101^*0)^*0)^*(01 + A(101^*0)^*11))^*$$

Problem 11. Create an automaton to recognize decimal numbers divisible by 5 and find the regular expression. The digits are read from left to right.

Answer. This problem is much easier than the binary divisibility by 5 problem. For a decimal number to be divisible by 5, the last digit must be a 0 or 5. The automaton for this is shown below.

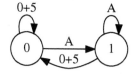

where 0 is the start and accept state and $A = 1 + 2 + 3 + 4 + 6 + 7 + 8 + 9$.

Problem 12. Create an automaton to recognize binary numbers divisible by 6 and find the regular expression. The bits are read from left to right.

Answer. A binary number that is divisible by 3 can be made divisible by 6, by simply multiplying it by 2. This is done by adding a 0 to the end. So we can start with the automaton for divisibility by 3 and modify it so that the number must end with a 0 to be accepted. The automaton for divisibility by 3 is reproduced here from page 26.

In the transition from state 1 to state 0, we insert a state 3 which goes to state 0 on a 0 following one or more 1's.

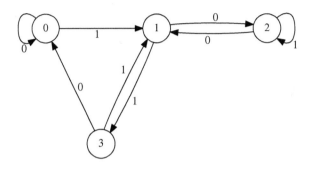

Now we eliminate state 2 which gives

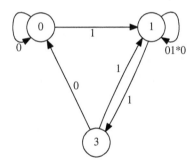

Then eliminating state 3 results in

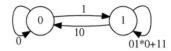

So the regular expression must be

$$R = (0 + 1(01^*0 + 11)^*10)^*$$

Problem 13. Show that the minimum number of states for testing if a binary number is divisible by $3 \cdot 2^n$ is $n + 3$.

Answer. This is similar to the division by 6 problem. The number must be divisible by 3 and then it must end in n zeros to make it also divisible by 2^n. Detecting the divisibility by 3 takes 3 states and then detecting the n zeros takes another n states for a total of $n + 3$ states.

Problem 14. Create an automaton to recognize binary numbers divisible by 7 and find the regular expression. The bits are read from left to right.

Answer. Let b be the value of the next bit, then the state will switch from r to r' according to the equation $r' = (2r + b) \bmod 7$. The table below shows how the states switch depending on the value of b. The table is used to construct the automaton shown in the figure below.

$$b$$

r	0	1
0	0	1
1	2	3
2	4	5
3	6	0
4	1	2
5	3	4
6	5	6

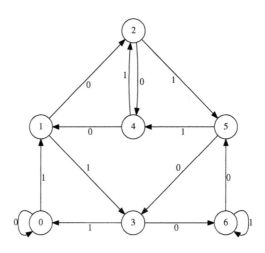

We begin simplifying it by removing state 2, which gives

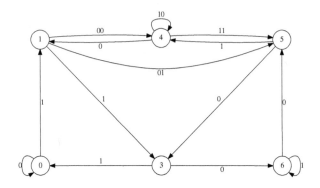

Then removing state 4 leaves

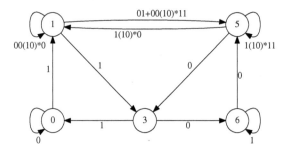

And removing state 3 gives

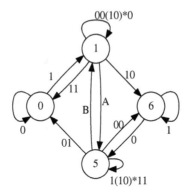

where $A = 01 + 00(10)^*11$ and $B = 1(10)^*0$.
Now we eliminate node 6 leaving

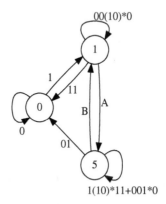

where $A = 01 + 00(10)^*11 + 101^*0$ and $B = 1(10)^*0$.

Finally, we remove node 5 to get

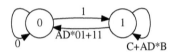

where $A = 01 + 00(10)^*11 + 101^*0$, $B = 1(10)^*0$, $C = 00(10)^*0$, and $D = 1(10)^*11 + 001^*0$.

From this, we can read off the regular expression as $R = (0 + 1(C + AD^*B)^*(AD^*01 + 11))^*$

and substituting in the values for A, B, C, and D gives

$R = (0 + 1(00(10)^*0 + (01 + 00(10)^*11 + 101^*0)$ $(1(10)^*11 + 001^*0)^*1(10)^*0)^*((01 + 00(10)^*11 + 101^*0)(1(10)^*11 + 001^*0)^*01 + 11))^*$

The 6-bit binary numbers that satisfy this regular expression, along with their base-10 equivalent are:

```
000000  0
000111  7
001110  14
```

```
010101  21
011100  28
100011  35
101010  42
110001  49
111000  56
111111  63
```

Problem 15. Create an automaton to recognize binary numbers divisible by 10.

Answer. To be divisible by 10 the number has to be divisible by 5 and 2. Divisibility by 2 means that the number must end with a zero. The automaton is then just a simple modification of the divisibility by 5 automaton. Only one state has to be added before the return to state 0 to detect a zero at the end of the number. The automaton is shown below.

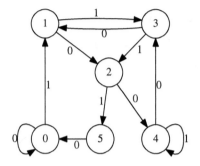

Patterns

Given a string constructed from the binary alphabet $\{a, b\}$ how do you use an automaton to search for the pattern aba? An obvious answer is to use an automaton with three states, in addition to the start state, that indicate how much of the pattern has been read so far. The automaton for doing this is

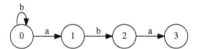

The corresponding regular expression is any number of b's followed by the pattern: $R = b^*aba$.

It works but it's very inefficient. The attempted match starts at the first character of the string and continues until the pattern is complete or there is a mismatch.

On a mismatch the process starts again at the second character of the string and so on. For a string like *abbaba* this process makes no sense because a match can not start at the second character if the first two characters matched. It is possible to do better.

One idea is to try to keep the automaton going so that it matches everything up to and including the pattern. In other words it should match what is not the pattern followed by the pattern. That way it doesn't have to stop and restart constantly. So whenever there is a mismatch we could put in a transition back to the start state (state 0). The automaton for doing this is

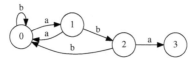

The corresponding regular expression is
$R = (b + aa + abb)^*aba$

It works well for the string *abbaba*. Everything is matched up to the pattern at the end so there is no stopping and restarting. But does it work for all strings? The answer is no. Given the string *abbaaba* you get the same problem. There is a mismatch when starting at each of the first four characters. Only the pattern at the end will match after restarting for the fifth time. Simply transitioning back to the start on a mismatch does not always work.

In this particular case the problem is state 1. Being in state 1 means the first character of the pattern, an a, has been read. If the next character is also an a then we should stay in state 1 and not go back to state 0. The automaton for doing this is

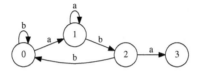

The corresponding regular expression is
$R = (b + aa^*bb)^*aa^*ba$

This automaton will work for any binary string, matching everything up to and including the pattern if it exists. If the pattern doesn't exist the automaton will continue to the end of the string and then fail because it doesn't end in the accepting state.

The question is how do you construct an automaton like this for arbitrary patterns and alphabets? The automaton only has to keep track of how many of the last characters read match the pattern. In other words it needs to remember the length of the longest prefix of the pattern that is a suffix of what has been read so far. If the length of the pattern is m then the automaton must have $m + 1$ states labeled from 0 to m. In state 0 no prefix of the pattern is a suffix of what has been read. In state m the entire pattern is a suffix of what has been read and the match is complete.

To analyze things further we need some notation. Let a pattern of length m be $p_{1,m} = p_1 p_2 \cdots p_m$ where p_i is the ith character of the pattern. A length k prefix of the pattern is $p_{1,k} = p_1 p_2 \cdots p_k$ and in general if $k > i$ then a length $k - i + 1$ section of the pattern is $p_{i,k} = p_i p_{i+1} \cdots p_k$.

Let the transition function for the automaton be $\delta(q_i, a) = q_j$, meaning that if the automaton is in state q_i and it reads character a then it goes to state q_j. We want the automaton to be in state q if the last q characters read equal $p_{1,q}$ and the last $q + k$ characters do not equal $p_{1,q+k}$ for any $k > 0$. For this to be true the transition function must be $\delta(0, p_1) = 1$, $\delta(1, p_2) = 2$, and in general $\delta(q, p_{q+1}) = q + 1$. But if we are in state q and the next character read is not p_{q+1}, where do we go?

The transitions for states 0 and 1 are obvious. For state 0, if $a \neq p_1$ then $\delta(0, a) = 0$. For state 1, if $a \neq p_2$ then $\delta(1, a) = 1$ when $a = p_1$ and $\delta(1, a) = 0$ when $a \neq p_1$. Now let's look at state 2. If a is read in state 2 then the last three characters read must be p_1p_2a. If $a = p_3$ the transition is to state 3. If $p_2a = p_1p_2$ (possible only if $a = p_2 = p_1$) the transition is to state 2. If $a = p_1$ the transition is to state 1. Otherwise the transition is to state 0.

These examples can easily be generalized. If the character a is read with the automaton in state q then the last $q + 1$ characters read must be $p_{1,q}a$.

If $p_{1,q}a = p_{1,q+1}$ the transition is to state $q + 1$.

If $p_{2,q}a = p_{1,q}$ the transition is to state q.

If $p_{3,q}a = p_{1,q-1}$ the transition is to state $q - 1$.

If $p_{q-k+2,q}a = p_{1,k}$ the transition is to state k.

In general when the next character read is a then the transition is to state k where k is the largest number such that $a = p_k$ and $p_{q-k+2,q} = p_{1,k-1}$. This simple rule can be used to design an automaton for matching any pattern. The accepting state is of course state m. To match strings that do not contain the pattern you simply make states 0 through $m - 1$ the accepting states.

Pattern Problems

Problem 16. Find a regular expression for recognizing binary numbers. Look at the case where leading zeros are allowed and where they are not.

Answer. Leading zeros allowed: $R = (0 + 1)^*$. Leading zeros not allowed: $R = \epsilon + 0 + 1(0 + 1)^*$

Problem 17. Find a regular expression for recognizing positive integers in base 10. Do not allow leading zeros.

Answer. $R = \epsilon + 0 + [1 - 9][0 - 9]^*$

Problem 18. Find a regular expression for recognizing real numbers in base 10. Allow one leading zero only immediately before the decimal point.

Answer. $R = (\epsilon + 0 + [1 - 9][0 - 9]^*).[0 - 9]^*$

Problem 19. Find a regular expression for recognizing binary numbers that contain exactly three ones. Create the corresponding automaton.

Answer. There can be any number of 0's at the beginning, at the end, and between the three 1's, so the regular expression is $R = 0^*10^*10^*10^*$. The automaton has three states corresponding to the number of 1's. The start state is 0 and the end state is 3.

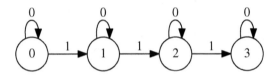

The 5-bit binary numbers that satisfy this regular expression, along with their base-10 equivalent are:

```
00111 7
01011 11
01101 13
01110 14
10011 19
10101 21
```

55

```
10110  22
11001  25
11010  26
11100  28
```

Problem 20. Given the alphabet $\Sigma = \{a, b, c, d\}$, find
a regular expression for recognizing words that
contain exactly one c and one d. Create the corresponding automaton.

Answer. There can be any number of a's or b's at the
beginning, at the end and between the c and d.
Also the c can come before or after the d so the
regular expression is $R = (a + b)^*(c(a + b)^*d +
d(a + b)^*c)(a + b)^*$. The automaton has 4 states.
The start state is 0 and the end state is 3. Note
that if the automaton is in state 1 and it reads a
c then there is no transition so it stops and the
word is not accepted. It also stops if it reads a d
in state 2 or a c or d in state 3.

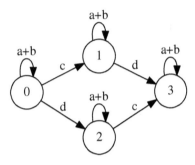

Problem 21. Same as previous problem except that the words should contain exactly two c's and two d's.

Answer. The words in this language are specified by regular expressions of the form: $(a + b)x(a + b)x(a + b)x(a + b)x(a + b)$ where two of the four x's will equal c and two will equal d. There will be $\binom{4}{2} = 6$ regular expressions of this form corresponding to ways values can be assigned to the x's.

Problem 22. Show how to generalize the previous two problems so that words containing exactly r copies of c and s copies of d are recognized. Describe how the regular expression is constructed.

Answer. The r c's and s d's can appear in $\binom{r+s}{r}$ different orders and there will have to be a regular expression for each order with any number of a's and b's at the beginning, at the end, and between the c's and d's.

Problem 23. Find a regular expression for recognizing binary numbers with no runs of three or more ones. Create the corresponding automaton.

Answer. There can be runs of zero, one, or two 1's followed by a 0 and this can be repeated any number of times. A final run of zero, one, or two 1's can appear at the end so the regular expression is
$R = ((\epsilon + 1 + 11)0)^*(\epsilon + 1 + 11)$
and the automaton is

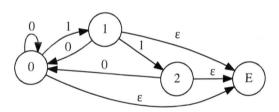

The start state is 0 and the end state is E.

The 4-bit binary numbers that satisfy this regular expression, along with their base-10 equivalent are:

```
0000  0
0001  1
0010  2
0011  3
0100  4
0101  5
0110  6
1000  8
1001  9
1010  10
1011  11
1100  12
1101  13
```

Problem 24. Generalize the previous problem to no runs of k or more ones.

Answer. Let $A = \epsilon + 1 + 1^2 + \cdots + 1^{k-1}$ be the regular expression for a run of length less than k, then, analogous to the previous problem, the regular expression will be $R = (A0)^* A$.

Problem 25. Find a regular expression for recognizing binary numbers with no runs of three or more ones or zeros.

Answer. The regular expression is
$$R = (\epsilon + 1 + 11)((0 + 00)(1 + 11))^*(\epsilon + 0 + 00)$$

The 4-bit binary numbers that satisfy this regular expression, along with their base-10 equivalent are:

```
0010 2
0011 3
0100 4
0101 5
0110 6
1001 9
1010 10
1011 11
1100 12
1101 13
```

Problem 26. Generalize the previous problem to no runs of k or more ones and l or more zeros.

Answer. The regular expression is
$$R = (\epsilon+1+1^2+\cdots+1^{k-1})((0+0^2+\cdots+0^{l-1})(1+ 1^2 + \cdots + 1^{k-1}))^*(\epsilon + 0 + 0^2 + \cdots + 0^{l-1})$$

Problem 27. Find a regular expression for recognizing binary numbers where ones only appear in runs of 2 or more.

Answer. The regular expression for getting zero or more runs of 2 or more is
$$R = (0 + 111^*0)^*(\epsilon + 111^*)$$
The regular expression for getting at least one run of 2 or more is
$$R = 0^*(111^*00^*)^*111^*0^*$$

The 5-bit binary numbers that satisfy the second regular expression, along with their base-10 equivalent are:

```
00011 3
00110 6
00111 7
01100 12
01110 14
01111 15
11000 24
11011 27
11100 28
11110 30
11111 31
```

Problem 28. Find a regular expression for recognizing binary numbers where both ones and zeros only appear in runs of 2 or more. Create the corresponding automaton.

Answer. The regular expression is
$$R = (000^* + 111^*)^*$$

The 6-bit binary numbers that satisfy this regular expression, along with their base-10 equivalent are:

```
000000 0
000011 3
000111 7
001100 12
001111 15
110000 48
110011 51
111000 56
111100 60
111111 63
```

Binary 3-Bit Patterns

Note that if we have the automaton and regular expression for a pattern such as 000, then the automaton and regular expression for the **NOT** of the pattern, in this case 111, is gotten by swapping 1's and 0's in the automaton and regular expression of the original pattern. Thus the number of unique patterns for an n bit binary string is $2^n/2$.

Problem 29. Construct an automaton that accepts binary words where the pattern 000 occurs only

at the end of the word. Find the corresponding regular expression and generating function.
Also find the regular expression and generating
function for binary words that do not contain the
pattern anywhere.

Answer.

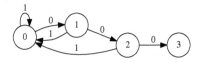

$$R = (1 + 0(1 + 01))^*000$$

$$G = -\frac{z^3}{z^3 + z^2 + z - 1}$$
$$= z^3 + z^4 + 2z^5 + 4z^6 + 7z^7 + 13z^8 + 24z^9$$
$$+ 44z^{10} + 81z^{11} + 149z^{12} + 274z^{13}$$
$$+ 504z^{14} + 927z^{15} + \cdots$$

The regular expression and generating function
for not finding the pattern anywhere is
$$R = (1 + 0(1 + 01))^*(00?)?$$

$$G = -\frac{z^2 + z + 1}{z^3 + z^2 + z - 1}$$
$$= 1 + 2z + 4z^2 + 7z^3 + 13z^4 + 24z^5 + 44z^6$$
$$+ 81z^7 + 149z^8 + 274z^9 + 504z^{10} + 927z^{11}$$
$$+ 1705z^{12} + 3136z^{13} + 5768z^{14}$$
$$+ 10609z^{15} + \cdots$$

Problem 30. Construct an automaton that accepts binary words where the pattern 000 occurs at the end of the word and anywhere before the end. Find the corresponding regular expression and generating function. Also find the regular expression and generating function for binary words that do not contain the pattern at the end.

Answer.

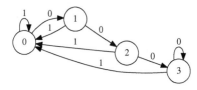

$$R = (1 + 0(1 + 0(1 + 00^*1)))^*0000^*$$

$$G = -\frac{z^3}{2z - 1}$$
$$= z^3 + 2z^4 + 4z^5 + 8z^6 + 16z^7 + 32z^8 + 64z^9$$
$$+ 128z^{10} + 256z^{11} + 512z^{12} + 1024z^{13}$$
$$+ 2048z^{14} + 4096z^{15} + \cdots$$

The regular expression and generating function for not finding the pattern at the end is
$R = (1 + 0(1 + 0(1 + 00^*1)))^*(00?)$?

$$G = \frac{z^3 - 1}{2z - 1}$$
$$= 1 + 2z + 4z^2 + 7z^3 + 14z^4 + 28z^5 + 56z^6$$
$$+ 112z^7 + 224z^8 + 448z^9 + 896z^{10} + 1792z^{11}$$
$$+ 3584z^{12} + 7168z^{13} + 14336z^{14} + 28672z^{15}$$
$$+ \cdots$$

Problem 31. Construct an automaton that accepts binary words where the pattern 001 occurs only at the end of the word. Find the corresponding regular expression and generating function. Also find the regular expression and generating function for binary words that do not contain the pattern anywhere.

Answer.

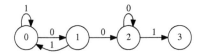

$R = (1 + 01)^{*}000^{*}1$

$$G = \frac{z^3}{z^3 - 2z + 1}$$
$$= z^3 + 2z^4 + 4z^5 + 7z^6 + 12z^7 + 20z^8$$
$$+ 33z^9 + 54z^{10} + 88z^{11} + 143z^{12} + 232z^{13}$$
$$+ 376z^{14} + 609z^{15} + \cdots$$

The regular expression and generating function for not finding the pattern anywhere is
$R = (1 + 01)^{*}(0(00^{*})?)$?

$$G = \frac{1}{z^3 - 2z + 1}$$
$$= 1 + 2z + 4z^2 + 7z^3 + 12z^4 + 20z^5 + 33z^6$$
$$+ 54z^7 + 88z^8 + 143z^9 + 232z^{10} + 376z^{11}$$
$$+ 609z^{12} + 986z^{13} + 1596z^{14} + 2583z^{15}$$
$$+ \cdots$$

Problem 32. Construct an automaton that accepts binary words where the pattern 001 occurs at the end of the word and anywhere before the end. Find the corresponding regular expression and generating function. Also find the regular expression and generating function for binary words that do not contain the pattern at the end.

Answer.

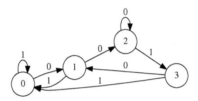

$$R = (1 + 0(00^*10)^*(1 + 00^*11))^*0(00^*10)^*00^*1$$

$$
\begin{aligned}
G &= -\frac{z^3}{2z - 1} \\
&= z^3 + 2z^4 + 4z^5 + 8z^6 + 16z^7 + 32z^8 + 64z^9 \\
&\quad + 128z^{10} + 256z^{11} + 512z^{12} + 1024z^{13} \\
&\quad + 2048z^{14} + 4096z^{15} + \cdots
\end{aligned}
$$

The regular expression and generating function for not finding the pattern at the end is
$R = (1+0(00^*10)^*(1+00^*11))^*(0(00^*10)^*(00^*)?)?$

$$G = \frac{z^3 - 1}{2z - 1}$$
$$= 1 + 2z + 4z^2 + 7z^3 + 14z^4 + 28z^5 + 56z^6$$
$$+ 112z^7 + 224z^8 + 448z^9 + 896z^{10} + 1792z^{11}$$
$$+ 3584z^{12} + 7168z^{13} + 14336z^{14} + 28672z^{15}$$
$$+ \cdots$$

Problem 33. Construct an automaton that accepts binary words where the pattern 010 occurs only at the end of the word. Find the corresponding regular expression and generating function. Also find the regular expression and generating function for binary words that do not contain the pattern anywhere.

Answer.

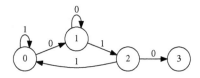

$$R = (1 + 00^*11)^*00^*10$$

$$G = -\frac{z^3}{z^3 - z^2 + 2z - 1}$$
$$= z^3 + 2z^4 + 3z^5 + 5z^6 + 9z^7 + 16z^8 + 28z^9$$
$$+ 49z^{10} + 86z^{11} + 151z^{12} + 265z^{13}$$
$$+ 465z^{14} + 816z^{15} + \cdots$$

The regular expression and generating function for not finding the pattern anywhere is $R = (1 + 00^*11)^*(00^*1?)$?

$$G = -\frac{z^2 + 1}{z^3 - z^2 + 2z - 1}$$
$$= 1 + 2z + 4z^2 + 7z^3 + 12z^4 + 21z^5 + 37z^6 + 65z^7$$
$$+ 114z^8 + 200z^9 + 351z^{10} + 616z^{11} + 1081z^{12}$$
$$+ 1897z^{13} + 3329z^{14} + 5842z^{15} + \cdots$$

Problem 34. Construct an automaton that accepts binary words where the pattern 010 occurs at the end of the word and anywhere before the end. Find the corresponding regular expression and generating function. Also find the regular expression and generating function for binary words that do not contain the pattern at the end.

Answer.

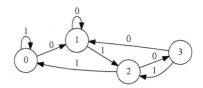

$R = (1+0(0+1(01)*00)*1(01)*1)*0(0+1(01)*00)*$
$1(01)*0$

$$G = -\frac{z^3}{2z-1}$$
$$= z^3 + 2z^4 + 4z^5 + 8z^6 + 16z^7 + 32z^8 + 64z^9$$
$$+ 128z^{10} + 256z^{11} + 512z^{12} + 1024z^{13}$$
$$+ 2048z^{14} + 4096z^{15} + \cdots$$

The regular expression and generating function
for not finding the pattern at the end is
$R = (1+0(0+1(01)*00)*1(01)*1)*(0(0+1(01)*00)*$
$(1(01)*)?)?$

$$G = \frac{z^3 - 1}{2z - 1}$$
$$= 1 + 2z + 4z^2 + 7z^3 + 14z^4 + 28z^5 + 56z^6$$
$$+ 112z^7 + 224z^8 + 448z^9 + 896z^{10} + 1792z^{11}$$
$$+ 3584z^{12} + 7168z^{13} + 14336z^{14} + 28672z^{15}$$
$$+ \cdots$$

Problem 35. Construct an automaton that accepts binary words where the pattern 011 occurs only at the end of the word. Find the corresponding regular expression and generating function. Also find the regular expression and generating function for binary words that do not contain the pattern anywhere.

Answer.

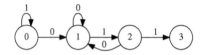

$R = 1^*0(0 + 10)^*11$

$$G = \frac{z^3}{z^3 - 2z + 1}$$
$$= z^3 + 2z^4 + 4z^5 + 7z^6 + 12z^7 + 20z^8 + 33z^9$$
$$+ 54z^{10} + 88z^{11} + 143z^{12} + 232z^{13}$$
$$+ 376z^{14} + 609z^{15} + \cdots$$

The regular expression and generating function for not finding the pattern anywhere is
$R = 1^*(0(0 + 10)^*1?)?$

$$G = \frac{1}{z^3 - 2z + 1}$$
$$= 1 + 2z + 4z^2 + 7z^3 + 12z^4 + 20z^5 + 33z^6$$
$$+ 54z^7 + 88z^8 + 143z^9 + 232z^{10} + 376z^{11}$$
$$+ 609z^{12} + 986z^{13} + 1596z^{14} + 2583z^{15} + \cdots$$

Problem 36. Construct an automaton that accepts binary words where the pattern 011 occurs at the end of the word and anywhere before the end. Find the corresponding regular expression and generating function. Also find the regular expression and generating function for binary words that do not contain the pattern at the end.

Answer.

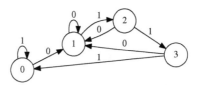

$$R = (1 + 0(0 + 1(0 + 10)) {}^*111) {}^*0(0 + 1(0 + 10)) {}^*11$$

$$G = -\frac{z^3}{2z - 1}$$
$$= z^3 + 2z^4 + 4z^5 + 8z^6 + 16z^7 + 32z^8 + 64z^9$$
$$+ 128z^{10} + 256z^{11} + 512z^{12} + 1024z^{13}$$
$$+ 2048z^{14} + 4096z^{15} + \cdots$$

The regular expression and generating function for not finding the pattern at the end is
$R = (1 + 0(0 + 1(0 + 10))^*111)^*(0(0 + 1(0 + 10))^*1?)$?

$$G = \frac{z^3 - 1}{2z - 1}$$
$$= 1 + 2z + 4z^2 + 7z^3 + 14z^4 + 28z^5 + 56z^6 + 112z^7$$
$$+ 224z^8 + 448z^9 + 896z^{10} + 1792z^{11} + 3584z^{12}$$
$$+ 7168z^{13} + 14336z^{14} + 28672z^{15} + \cdots$$

Binary 4-Bit Patterns

Problem 37. Construct an automaton that accepts binary words where the pattern 0000 occurs only at the end of the word. Find the corresponding regular expression and generating function. Also find the regular expression and generating function for binary words that do not contain the pattern anywhere.

Answer.

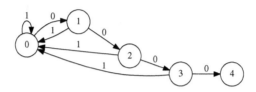

$R = (1 + 0(1 + 0(1 + 01)))^*0000$

$$G = -\frac{z^4}{z^4 + z^3 + z^2 + z - 1}$$
$$= z^4 + z^5 + 2z^6 + 4z^7 + 8z^8 + 15z^9 + 29z^{10}$$
$$+ 56z^{11} + 108z^{12} + 208z^{13} + 401z^{14}$$
$$+ 773z^{15} + \cdots$$

The regular expression and generating function
for not finding the pattern anywhere is
$R = (1 + 0(1 + 0(1 + 01)))^*(0(00?)?)?$

$$G = -\frac{z^3 + z^2 + z + 1}{z^4 + z^3 + z^2 + z - 1}$$
$$= 1 + 2z + 4z^2 + 8z^3 + 15z^4 + 29z^5 + 56z^6$$
$$+ 108z^7 + 208z^8 + 401z^9 + 773z^{10} + 1490z^{11}$$
$$+ 2872z^{12} + 5536z^{13} + 10671z^{14}$$
$$+ 20569z^{15} + \cdots$$

Problem 38. Construct an automaton that accepts binary words where the pattern 0000 occurs at the end of the word and anywhere before the end. Find the corresponding regular expression and generating function. Also find the regular expression and generating function for binary words that do not contain the pattern at the end.

Answer.

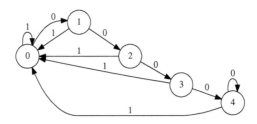

$$R = (1 + 0(1 + 0(1 + 0(1 + 00^*1))))^*00000^*$$

$$G = -\frac{z^4}{2z - 1}$$
$$= z^4 + 2z^5 + 4z^6 + 8z^7 + 16z^8 + 32z^9 + 64z^{10}$$
$$+ 128z^{11} + 256z^{12} + 512z^{13} + 1024z^{14}$$
$$+ 2048z^{15} + \cdots$$

The regular expression and generating function for not finding the pattern at the end is

$$R = (1 + 0(1 + 0(1 + 0(1 + 00^*1))))^*(0(00?)?)?$$

$$G = \frac{z^4 - 1}{2z - 1}$$
$$= 1 + 2z + 4z^2 + 8z^3 + 15z^4 + 30z^5 + 60z^6$$
$$+ 120z^7 + 240z^8 + 480z^9 + 960z^{10} + 1920z^{11}$$
$$+ 3840z^{12} + 7680z^{13} + 15360z^{14}$$
$$+ 30720z^{15} + \cdots$$

Problem 39. Construct an automaton that accepts binary words where the pattern 0001 occurs only at the end of the word. Find the corresponding regular expression and generating function. Also find the regular expression and generating function for binary words that do not contain the pattern anywhere.

Answer.

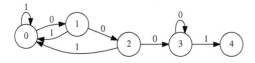

$$R = (1 + 0(1 + 01))^*0000^*1$$

$$G = \frac{z^4}{z^4 - 2z + 1}$$
$$= z^4 + 2z^5 + 4z^6 + 8z^7 + 15z^8 + 28z^9 + 52z^{10}$$
$$+ 96z^{11} + 177z^{12} + 326z^{13} + 600z^{14}$$
$$+ 1104z^{15} + \cdots$$

The regular expression and generating function for not finding the pattern anywhere is
$$R = (1 + 0(1 + 01))^*(0(0(00^*)?)?)?$$

$$G = \frac{1}{z^4 - 2z + 1}$$
$$= 1 + 2z + 4z^2 + 8z^3 + 15z^4 + 28z^5 + 52z^6 + 96z^7$$
$$+ 177z^8 + 326z^9 + 600z^{10} + 1104z^{11} + 2031z^{12}$$
$$+ 3736z^{13} + 6872z^{14} + 12640z^{15} + \cdots$$

Problem 40. Construct an automaton that accepts binary words where the pattern 0001 occurs at the end of the word and anywhere before the end. Find the corresponding regular expression and generating function. Also find the regular expression and generating function for binary words that do not contain the pattern at the end.

Answer.

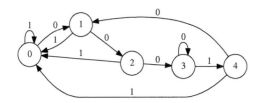

$R = (1+0(000^*10)^*(1+0(1+00^*11)))^*0(000^*10)^*$
000^*1

$$G = -\frac{z^4}{2z - 1}$$
$$= z^4 + 2z^5 + 4z^6 + 8z^7 + 16z^8 + 32z^9 + 64z^{10}$$
$$+ 128z^{11} + 256z^{12} + 512z^{13} + 1024z^{14}$$
$$+ 2048z^{15} + \cdots$$

The regular expression and generating function
for not finding the pattern at the end is
$R = (1+0(000^*10)^*(1+0(1+00^*11)))^*(0(000^*10)^*$
$(0(00^*)?)?)?$

$$G = \frac{z^4 - 1}{2z - 1}$$
$$= 1 + 2z + 4z^2 + 8z^3 + 15z^4 + 30z^5 + 60z^6$$
$$+ 120z^7 + 240z^8 + 480z^9 + 960z^{10} + 1920z^{11}$$
$$+ 3840z^{12} + 7680z^{13} + 15360z^{14}$$
$$+ 30720z^{15} + \cdots$$

Problem 41. Construct an automaton that accepts
binary words where the pattern 0010 occurs only
at the end of the word. Find the corresponding regular expression and generating function.
Also find the regular expression and generating
function for binary words that do not contain the
pattern anywhere.

Answer.

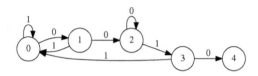

$$R = (1 + 0(1 + 00^*11))^*000^*10$$

$$G = -\frac{z^4}{z^4 - z^3 + 2z - 1}$$
$$= z^4 + 2z^5 + 4z^6 + 7z^7 + 13z^8 + 24z^9$$
$$+ 45z^{10} + 84z^{11} + 157z^{12} + 293z^{13}$$
$$+ 547z^{14} + 1021z^{15} + \cdots$$

The regular expression and generating function
for not finding the pattern anywhere is

$$R = (1 + 0(1 + 00^*11))^*(0(00^*1?)?)?$$

$$G = -\frac{z^3 + 1}{z^4 - z^3 + 2z - 1}$$
$$= 1 + 2z + 4z^2 + 8z^3 + 15z^4 + 28z^5 + 52z^6$$
$$+ 97z^7 + 181z^8 + 338z^9 + 631z^{10} + 1178z^{11}$$
$$+ 2199z^{12} + 4105z^{13} + 7663z^{14}$$
$$+ 14305z^{15} + \cdots$$

Problem 42. Construct an automaton that accepts binary words where the pattern 0010 occurs at the end of the word and anywhere before the end. Find the corresponding regular expression and generating function. Also find the regular expression and generating function for binary words that do not contain the pattern at the end.

Answer.

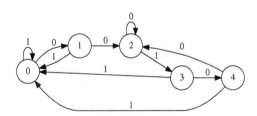

$$R = (1+0(1+0(0+100)\text{*}1(1+01)))\text{*}00(0+100)\text{*}10$$

$$G = -\frac{z^4}{2z-1}$$
$$= z^4 + 2z^5 + 4z^6 + 8z^7 + 16z^8 + 32z^9 + 64z^{10}$$
$$+ 128z^{11} + 256z^{12} + 512z^{13} + 1024z^{14}$$
$$+ 2048z^{15} + \cdots$$

The regular expression and generating function for not finding the pattern at the end is
$R = (1 + 0(1 + 0(0 + 100)\text{*}1(1 + 01)))\text{*}(0(0(0 + 100)\text{*}1?)?)?$

$$G = \frac{z^4 - 1}{2z - 1}$$
$$= 1 + 2z + 4z^2 + 8z^3 + 15z^4 + 30z^5 + 60z^6 + 120z^7$$
$$+ 240z^8 + 480z^9 + 960z^{10} + 1920z^{11} + 3840z^{12}$$
$$+ 7680z^{13} + 15360z^{14} + 30720z^{15} + \cdots$$

Problem 43. Construct an automaton that accepts binary words where the pattern 0011 occurs only at the end of the word. Find the corresponding regular expression and generating function. Also find the regular expression and generating function for binary words that do not contain the pattern anywhere.

Answer.

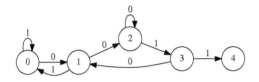

$$R = (1 + 0(00*10)*1)*0(00*10)*00*11$$

$$G = \frac{z^4}{z^4 - 2z + 1}$$
$$= z^4 + 2z^5 + 4z^6 + 8z^7 + 15z^8 + 28z^9 + 52z^{10}$$
$$+ 96z^{11} + 177z^{12} + 326z^{13} + 600z^{14}$$
$$+ 1104z^{15} + \cdots$$

The regular expression and generating function for not finding the pattern anywhere is
$$R = (1 + 0(00*10)*1)*(0(00*10)*(00*1?)?)?$$

$$G = \frac{1}{z^4 - 2z + 1}$$
$$= 1 + 2z + 4z^2 + 8z^3 + 15z^4 + 28z^5 + 52z^6$$
$$+ 96z^7 + 177z^8 + 326z^9 + 600z^{10} + 1104z^{11}$$
$$+ 2031z^{12} + 3736z^{13} + 6872z^{14}$$
$$+ 12640z^{15} + \cdots$$

Problem 44. Construct an automaton that accepts binary words where the pattern 0011 occurs at

the end of the word and anywhere before the
end. Find the corresponding regular expression
and generating function. Also find the regular expression and generating function for binary words
that do not contain the pattern at the end.

Answer.

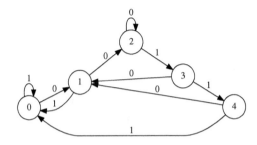

$$R = (1+0(00^*1(0+10))^*(1+00^*111))^*0(00^*1(0+ 10))^*00^*11$$

$$G = -\frac{z^4}{2z-1}$$
$$= z^4 + 2z^5 + 4z^6 + 8z^7 + 16z^8 + 32z^9 + 64z^{10}$$
$$+ 128z^{11} + 256z^{12} + 512z^{13} + 1024z^{14}$$
$$+ 2048z^{15} + \cdots$$

The regular expression and generating function
for not finding the pattern at the end is

$$R = (1+0(00^*1(0+10))^*(1+00^*111))^*(0(00^*1(0+10))^*(00^*1?)?)?$$

$$G = \frac{z^4 - 1}{2z - 1}$$
$$= 1 + 2z + 4z^2 + 8z^3 + 15z^4 + 30z^5 + 60z^6$$
$$+ 120z^7 + 240z^8 + 480z^9 + 960z^{10} + 1920z^{11}$$
$$+ 3840z^{12} + 7680z^{13} + 15360z^{14}$$
$$+ 30720z^{15} + \cdots$$

Problem 45. Construct an automaton that accepts binary words where the pattern 0100 occurs only at the end of the word. Find the corresponding regular expression and generating function. Also find the regular expression and generating function for binary words that do not contain the pattern anywhere.

Answer.

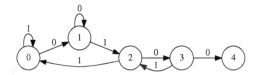

$$R = (1 + 00^*1(01)^*1)^*00^*1(01)^*00$$

$$G = -\frac{z^4}{z^4 - z^3 + 2z - 1}$$
$$= z^4 + 2z^5 + 4z^6 + 7z^7 + 13z^8 + 24z^9 + 45z^{10}$$
$$+ 84z^{11} + 157z^{12} + 293z^{13} + 547z^{14}$$
$$+ 1021z^{15} + \cdots$$

The regular expression and generating function
for not finding the pattern anywhere is
$$R = (1 + 00^*1(01)^*1)^*(00^*(1(01)^*0?)?)?$$

$$G = -\frac{z^3 + 1}{z^4 - z^3 + 2z - 1}$$
$$= 1 + 2z + 4z^2 + 8z^3 + 15z^4 + 28z^5 + 52z^6 + 97z^7$$
$$+ 181z^8 + 338z^9 + 631z^{10} + 1178z^{11} + 2199z^{12}$$
$$+ 4105z^{13} + 7663z^{14} + 14305z^{15} + \cdots$$

Problem 46. Construct an automaton that accepts
binary words where the pattern 0100 occurs at
the end of the word and anywhere before the
end. Find the corresponding regular expression
and generating function. Also find the regular expression and generating function for binary words
that do not contain the pattern at the end.

Answer.

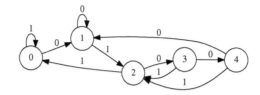

$R = (1+0(0+1(0(1+01))^*000)^*1(0(1+01))^*1)^*0(0+$
$1(0(1 + 01))^*000)^*1(0(1 + 01))^*00$

$$G = -\frac{z^4}{2z - 1}$$
$$= z^4 + 2z^5 + 4z^6 + 8z^7 + 16z^8 + 32z^9 + 64z^{10}$$
$$+ 128z^{11} + 256z^{12} + 512z^{13} + 1024z^{14}$$
$$+ 2048z^{15} + \cdots$$

The regular expression and generating function
for not finding the pattern at the end is
$R = (1+0(0+1(0(1+01))^*000)^*1(0(1+01))^*1)^*(0(0+$
$1(0(1 + 01))^*000)^*(1(0(1 + 01))^*0?)?)?$

$$G = \frac{z^4 - 1}{2z - 1}$$
$$= 1 + 2z + 4z^2 + 8z^3 + 15z^4 + 30z^5 + 60z^6$$
$$+ 120z^7 + 240z^8 + 480z^9 + 960z^{10} + 1920z^{11}$$
$$+ 3840z^{12} + 7680z^{13} + 15360z^{14}$$
$$+ 30720z^{15} + \cdots$$

Problem 47. Construct an automaton that accepts binary words where the pattern 0101 occurs only at the end of the word. Find the corresponding regular expression and generating function. Also find the regular expression and generating function for binary words that do not contain the pattern anywhere.

Answer.

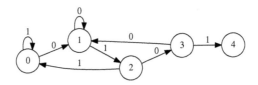

$$R = (1 + 0(0 + 100)^*11)^*0(0 + 100)^*101$$

$$G = \frac{z^4}{z^4 - 2z^3 + z^2 - 2z + 1}$$
$$= z^4 + 2z^5 + 3z^6 + 6z^7 + 12z^8 + 22z^9 + 41z^{10}$$
$$+ 78z^{11} + 147z^{12} + 276z^{13} + 520z^{14}$$
$$+ 980z^{15} + \cdots$$

The regular expression and generating function for not finding the pattern anywhere is

$$R = (1 + 0(0 + 100)^*11)^*(0(0 + 100)^*(10?)?)?$$

$$G = \frac{z^2 + 1}{z^4 - 2z^3 + z^2 - 2z + 1}$$
$$= 1 + 2z + 4z^2 + 8z^3 + 15z^4 + 28z^5 + 53z^6$$
$$+ 100z^7 + 188z^8 + 354z^9 + 667z^{10} + 1256z^{11}$$
$$+ 2365z^{12} + 4454z^{13} + 8388z^{14}$$
$$+ 15796z^{15} + \cdots$$

Problem 48. Construct an automaton that accepts binary words where the pattern 0101 occurs at the end of the word and anywhere before the end. Find the corresponding regular expression and generating function. Also find the regular expression and generating function for binary words that do not contain the pattern at the end.

Answer.

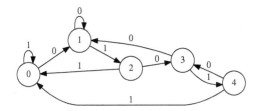

$R = (1 + 0(0 + 10(10)^*0)^*1(1 + 0(10)^*11))^*0(0 + 10(10)^*0)^*10(10)^*1$

$$G = -\frac{z^4}{2z - 1}$$
$$= z^4 + 2z^5 + 4z^6 + 8z^7 + 16z^8 + 32z^9 + 64z^{10}$$
$$+ 128z^{11} + 256z^{12} + 512z^{13} + 1024z^{14}$$
$$+ 2048z^{15} + \cdots$$

The regular expression and generating function for not finding the pattern at the end is
$R = (1 + 0(0 + 10(10)^*0)^*1(1 + 0(10)^*11))^*(0(0 + 10(10)^*0)^*(1(0(10)^*)?)?)?$

$$G = \frac{z^4 - 1}{2z - 1}$$
$$= 1 + 2z + 4z^2 + 8z^3 + 15z^4 + 30z^5 + 60z^6 + 120z^7$$
$$+ 240z^8 + 480z^9 + 960z^{10} + 1920z^{11} + 3840z^{12}$$
$$+ 7680z^{13} + 15360z^{14} + 30720z^{15} + \cdots$$

Problem 49. Construct an automaton that accepts binary words where the pattern 0110 occurs only at the end of the word. Find the corresponding regular expression and generating function. Also find the regular expression and generating function for binary words that do not contain the pattern anywhere.

Answer.

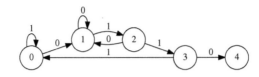

$$R = (1 + 0(0 + 10)^*111)^*0(0 + 10)^*110$$

$$G = -\frac{z^4}{z^4 - z^3 + 2z - 1}$$
$$= z^4 + 2z^5 + 4z^6 + 7z^7 + 13z^8 + 24z^9 + 45z^{10}$$
$$+ 84z^{11} + 157z^{12} + 293z^{13} + 547z^{14}$$
$$+ 1021z^{15} + \cdots$$

The regular expression and generating function
for not finding the pattern anywhere is
$$R = (1 + 0(0 + 10)^*111)^*(0(0 + 10)^*(11?)?)?$$

$$G = -\frac{z^3 + 1}{z^4 - z^3 + 2z - 1}$$
$$= 1 + 2z + 4z^2 + 8z^3 + 15z^4 + 28z^5 + 52z^6$$
$$+ 97z^7 + 181z^8 + 338z^9 + 631z^{10} + 1178z^{11}$$
$$+ 2199z^{12} + 4105z^{13} + 7663z^{14}$$
$$+ 14305z^{15} + \cdots$$

Problem 50. Construct an automaton that accepts binary words where the pattern 0110 occurs at the end of the word and anywhere before the end. Find the corresponding regular expression and generating function. Also find the regular expression and generating function for binary words that do not contain the pattern at the end.

Answer.

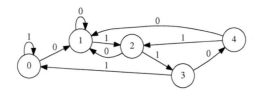

$$R = (1+0(0+1(101)^*(0+100))^*1(101)^*11)^*0(0+1(101)^*(0+100))^*1(101)^*10$$

$$
\begin{aligned}
G &= -\frac{z^4}{2z-1} \\
&= z^4 + 2z^5 + 4z^6 + 8z^7 + 16z^8 + 32z^9 + 64z^{10} \\
&\quad + 128z^{11} + 256z^{12} + 512z^{13} + 1024z^{14} \\
&\quad + 2048z^{15} + \cdots
\end{aligned}
$$

The regular expression and generating function for not finding the pattern at the end is

$$R = (1+0(0+1(101)^*(0+100))^*1(101)^*11)^*(0(0+$$
$$1(101)^*(0+100))^*(1(101)^*1?)?)?$$

$$G = \frac{z^4 - 1}{2z - 1}$$
$$= 1 + 2z + 4z^2 + 8z^3 + 15z^4 + 30z^5 + 60z^6$$
$$+ 120z^7 + 240z^8 + 480z^9 + 960z^{10} + 1920z^{11}$$
$$+ 3840z^{12} + 7680z^{13} + 15360z^{14}$$
$$+ 30720z^{15} + \cdots$$

Problem 51. Construct an automaton that accepts binary words where the pattern 0111 occurs only at the end of the word. Find the corresponding regular expression and generating function. Also find the regular expression and generating function for binary words that do not contain the pattern anywhere.

Answer.

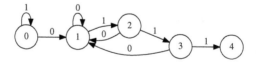

$$R = 1^*0(0 + 1(0 + 10))^*111$$

$$G = \frac{z^4}{z^4 - 2z + 1}$$
$$= z^4 + 2z^5 + 4z^6 + 8z^7 + 15z^8 + 28z^9 + 52z^{10}$$
$$+ 96z^{11} + 177z^{12} + 326z^{13} + 600z^{14}$$
$$+ 1104z^{15} + \cdots$$

The regular expression and generating function for not finding the pattern anywhere is
$$R = 1^*(0(0 + 1(0 + 10))^*(11?)?)?$$

$$G = \frac{1}{z^4 - 2z + 1}$$
$$= 1 + 2z + 4z^2 + 8z^3 + 15z^4 + 28z^5 + 52z^6 + 96z^7$$
$$+ 177z^8 + 326z^9 + 600z^{10} + 1104z^{11} + 2031z^{12}$$
$$+ 3736z^{13} + 6872z^{14} + 12640z^{15} + \cdots$$

Problem 52. Construct an automaton that accepts binary words where the pattern 0111 occurs at the end of the word and anywhere before the end. Find the corresponding regular expression and generating function. Also find the regular expression and generating function for binary words that do not contain the pattern at the end.

Answer.

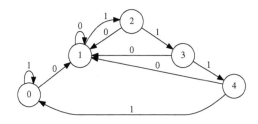

$R = (1+0(0+1(0+1(0+10)))^*1111)^*0(0+1(0+1(0+10)))^*111$

$$G = -\frac{z^4}{2z - 1}$$
$$= z^4 + 2z^5 + 4z^6 + 8z^7 + 16z^8 + 32z^9 + 64z^{10}$$
$$+ 128z^{11} + 256z^{12} + 512z^{13} + 1024z^{14}$$
$$+ 2048z^{15} + \cdots$$

The regular expression and generating function
for not finding the pattern at the end is
$R = (1 + 0(0 + 1(0 + 1(0 + 10)))^*1111)^*(0(0 + 1(0 + 1(0 + 10)))^*(11?)?)?$

$$G = \frac{z^4 - 1}{2z - 1}$$
$$= 1 + 2z + 4z^2 + 8z^3 + 15z^4 + 30z^5 + 60z^6$$
$$+ 120z^7 + 240z^8 + 480z^9 + 960z^{10} + 1920z^{11}$$
$$+ 3840z^{12} + 7680z^{13} + 15360z^{14}$$
$$+ 30720z^{15} + \cdots$$

Ternary Patterns

Problem 53. Construct an automaton that accepts ternary words (with alphabet 0,1,2) where the pattern 01 occurs only at the end of the word. Find the corresponding regular expression and generating function. Also find the regular expression and generating function for ternary words that do not contain the pattern anywhere.

Answer.

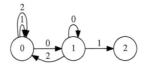

$$R = (1 + 2 + 00^*2)^*00^*1$$

$$G = \frac{z^2}{z^2 - 3z + 1}$$
$$= z^2 + 3z^3 + 8z^4 + 21z^5 + 55z^6 + 144z^7$$
$$+ 377z^8 + 987z^9 + 2584z^{10} + 6765z^{11}$$
$$+ 17711z^{12} + 46368z^{13} + 121393z^{14}$$
$$+ 317811z^{15} + \cdots$$

The regular expression and generating function for not finding the pattern anywhere is $R = (1 + 2 + 00^*2)^*(00^*)$?

$$G = \frac{1}{z^2 - 3z + 1}$$
$$= 1 + 3z + 8z^2 + 21z^3 + 55z^4 + 144z^5$$
$$+ 377z^6 + 987z^7 + 2584z^8 + 6765z^9$$
$$+ 17711z^{10} + 46368z^{11} + 121393z^{12}$$
$$+ 317811z^{13} + 832040z^{14} + 2178309z^{15} + \cdots$$

Problem 54. Construct an automaton that accepts ternary words where the pattern 01 occurs at the end of the word and anywhere before the end. Find the corresponding regular expression and generating function. Also find the regular expression and generating function for ternary words that do not contain the pattern at the end.

Answer.

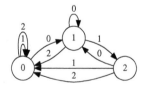

$$R = (1+2+0(0+10)^*(2+1(1+2)))^*0(0+10)^*1$$

$$G = -\frac{z^2}{3z-1}$$
$$= z^2 + 3z^3 + 9z^4 + 27z^5 + 81z^6 + 243z^7 + 729z^8$$
$$+ 2187z^9 + 6561z^{10} + 19683z^{11} + 59049z^{12}$$
$$+ 177147z^{13} + 531441z^{14} + 1594323z^{15} + \cdots$$

The regular expression and generating function for not finding the pattern at the end is
$$R = (1+2+0(0+10)^*(2+1(1+2)))^*(0(0+10)^*)?$$

$$G = \frac{z^2 - 1}{3z - 1}$$
$$= 1 + 3z + 8z^2 + 24z^3 + 72z^4 + 216z^5 + 648z^6$$
$$+ 1944z^7 + 5832z^8 + 17496z^9 + 52488z^{10}$$
$$+ 157464z^{11} + 472392z^{12} + 1417176z^{13}$$
$$+ 4251528z^{14} + 12754584z^{15} + \cdots$$

Problem 55. Construct an automaton that accepts ternary words where the pattern 012 occurs only at the end of the word. Find the corresponding regular expression and generating function. Also find the regular expression and generating function for ternary words that do not contain the pattern anywhere.

Answer.

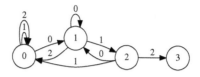

$$R = (1 + 2 + 0(0 + 10)^*(2 + 11))^*0(0 + 10)^*12$$

$$G = \frac{z^3}{z^3 - 3z + 1}$$
$$= z^3 + 3z^4 + 9z^5 + 26z^6 + 75z^7 + 216z^8$$
$$+ 622z^9 + 1791z^{10} + 5157z^{11} + 14849z^{12}$$
$$+ 42756z^{13} + 123111z^{14} + 354484z^{15} + \cdots$$

The regular expression and generating function for not finding the pattern anywhere is
$$R = (1 + 2 + 0(0 + 10)^*(2 + 11))^*(0(0 + 10)^*1?)?$$

$$G = \frac{1}{z^3 - 3z + 1}$$
$$= 1 + 3z + 9z^2 + 26z^3 + 75z^4 + 216z^5 + 622z^6$$
$$+ 1791z^7 + 5157z^8 + 14849z^9 + 42756z^{10}$$
$$+ 123111z^{11} + 354484z^{12} + 1020696z^{13}$$
$$+ 2938977z^{14} + 8462447z^{15} + \cdots$$

Problem 56. Construct an automaton that accepts ternary words where the pattern 012 occurs at the end of the word and anywhere before the end. Find the corresponding regular expression and generating function. Also find the regular expression and generating function for ternary words that do not contain the pattern at the end.

Answer.

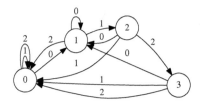

$$R = (1 + 2 + 0(0 + 1(0 + 20)))^*(2 + 1(1 + 2(1 + 2))))^*0(0 + 1(0 + 20))^*12$$

$$G = -\frac{z^3}{3z - 1}$$
$$= z^3 + 3z^4 + 9z^5 + 27z^6 + 81z^7 + 243z^8$$
$$+ 729z^9 + 2187z^{10} + 6561z^{11} + 19683z^{12}$$
$$+ 59049z^{13} + 177147z^{14} + 531441z^{15} + \cdots$$

The regular expression and generating function for not finding the pattern at the end is
$$R = (1 + 2 + 0(0 + 1(0 + 20)))^*(2 + 1(1 + 2(1 +$$

2))))*(0(0 + 1(0 + 20))*1?)?

$$
\begin{aligned}
G &= \frac{z^3 - 1}{3z - 1} \\
&= 1 + 3z + 9z^2 + 26z^3 + 78z^4 + 234z^5 + 702z^6 \\
&\quad + 2106z^7 + 6318z^8 + 18954z^9 + 56862z^{10} \\
&\quad + 170586z^{11} + 511758z^{12} + 1535274z^{13} \\
&\quad + 4605822z^{14} + 13817466z^{15} + \cdots
\end{aligned}
$$

Problem 57. Construct an automaton that accepts ternary words where the pattern 121 occurs only at the end of the word. Find the corresponding regular expression and generating function. Also find the regular expression and generating function for ternary words that do not contain the pattern anywhere.

Answer.

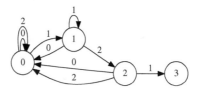

$$R = (0 + 2 + 11^*(0 + 2(0 + 2)))^*11^*21$$

$$G = -\frac{z^3}{2z^3 - z^2 + 3z - 1}$$
$$= z^3 + 3z^4 + 8z^5 + 23z^6 + 67z^7 + 194z^8 + 561z^9$$
$$+ 1623z^{10} + 4696z^{11} + 13587z^{12} + 39311z^{13}$$
$$+ 113738z^{14} + 329077z^{15} + \cdots$$

The regular expression and generating function for not finding the pattern anywhere is
$$R = (0 + 2 + 11^*(0 + 2(0 + 2)))^*(11^*2?)?$$

$$G = -\frac{z^2 + 1}{2z^3 - z^2 + 3z - 1}$$
$$= 1 + 3z + 9z^2 + 26z^3 + 75z^4 + 217z^5 + 628z^6$$
$$+ 1817z^7 + 5257z^8 + 15210z^9 + 44007z^{10}$$
$$+ 127325z^{11} + 368388z^{12} + 1065853z^{13}$$
$$+ 3083821z^{14} + 8922386z^{15} + \cdots$$

Problem 58. Construct an automaton that accepts ternary words where the pattern 121 occurs at the end of the word and anywhere before the end. Find the corresponding regular expression and generating function. Also find the regular expression and generating function for ternary words that do not contain the pattern at the end.

Answer.

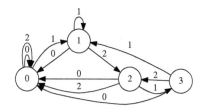

$$R = (0 + 2 + 1(1 + 2(12)^*11)^*(0 + 2(12)^*(0 + 2 + 10)))^*1(1 + 2(12)^*11)^*2(12)^*1$$

$$G = -\frac{z^3}{3z - 1}$$
$$= z^3 + 3z^4 + 9z^5 + 27z^6 + 81z^7 + 243z^8$$
$$+ 729z^9 + 2187z^{10} + 6561z^{11} + 19683z^{12}$$
$$+ 59049z^{13} + 177147z^{14} + 531441z^{15} + \cdots$$

The regular expression and generating function for not finding the pattern at the end is
$$R = (0 + 2 + 1(1 + 2(12)^*11)^*(0 + 2(12)^*(0 + 2 + 10)))^*(1(1 + 2(12)^*11)^*(2(12)^*)?)?$$

$$G = \frac{z^3 - 1}{3z - 1}$$
$$= 1 + 3z + 9z^2 + 26z^3 + 78z^4 + 234z^5 + 702z^6$$
$$+ 2106z^7 + 6318z^8 + 18954z^9 + 56862z^{10}$$
$$+ 170586z^{11} + 511758z^{12} + 1535274z^{13}$$
$$+ 4605822z^{14} + 13817466z^{15} + \cdots$$

Quaternary Patterns

Problem 59. Construct an automaton that accepts quaternary words (with alphabet 0,1,2,3) where the pattern 01 occurs only at the end of the word. Find the corresponding regular expression and generating function. Also find the regular expression and generating function for quaternary words that do not contain the pattern anywhere.

Answer.

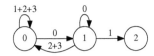

$$R = (1 + 2 + 3 + 00^*(2 + 3))^*00^*1$$

$$
\begin{aligned}
G &= \frac{z^2}{z^2 - 4z + 1} \\
&= z^2 + 4z^3 + 15z^4 + 56z^5 + 209z^6 + 780z^7 \\
&\quad + 2911z^8 + 10864z^9 + 40545z^{10} + 151316z^{11} \\
&\quad + 564719z^{12} + 2107560z^{13} + 7865521z^{14} \\
&\quad + 29354524z^{15} + \cdots
\end{aligned}
$$

The regular expression and generating function for not finding the pattern anywhere is $R = (1 + 2 + 3 + 00^*(2 + 3))^*(00^*)$?

$$G = \frac{1}{z^2 - 4z + 1}$$
$$= 1 + 4z + 15z^2 + 56z^3 + 209z^4 + 780z^5$$
$$+ 2911z^6 + 10864z^7 + 40545z^8 + 151316z^9$$
$$+ 564719z^{10} + 2107560z^{11} + 7865521z^{12}$$
$$+ 29354524z^{13} + 109552575z^{14}$$
$$+ 408855776z^{15} + \cdots$$

Problem 60. Construct an automaton that accepts quaternary words where the pattern 01 occurs at the end of the word and anywhere before the end. Find the corresponding regular expression and generating function. Also find the regular expression and generating function for quaternary words that do not contain the pattern at the end.

Answer.

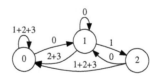

$$R = (1 + 2 + 3 + 0(0 + 10)^*(2 + 3 + 1(1 + 2 + 3)))^*0(0 + 10)^*1$$

$$G = -\frac{z^2}{4z - 1}$$
$$= z^2 + 4z^3 + 16z^4 + 64z^5 + 256z^6 + 1024z^7$$
$$+ 4096z^8 + 16384z^9 + 65536z^{10} + 262144z^{11}$$
$$+ 1048576z^{12} + 4194304z^{13} + 16777216z^{14}$$
$$+ 67108864z^{15} + \cdots$$

The regular expression and generating function for not finding the pattern at the end is
$$R = (1 + 2 + 3 + 0(0 + 10)^*(2 + 3 + 1(1 + 2 + 3)))^*(0(0 + 10)^*)?$$

$$G = \frac{z^2 - 1}{4z - 1}$$
$$= 1 + 4z + 15z^2 + 60z^3 + 240z^4 + 960z^5$$
$$+ 3840z^6 + 15360z^7 + 61440z^8 + 245760z^9$$
$$+ 983040z^{10} + 3932160z^{11} + 15728640z^{12}$$
$$+ 62914560z^{13} + 251658240z^{14}$$
$$+ 1006632960z^{15} + \cdots$$

Problem 61. Construct an automaton that accepts quaternary words where the pattern 131 occurs only at the end of the word. Find the corresponding regular expression and generating function.

Also find the regular expression and generating function for quaternary words that do not contain the pattern anywhere.

Answer.

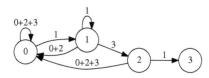

$$R = (0+2+3+11^*(0+2+3(0+2+3)))^*11^*31$$

$$G = -\frac{z^3}{3z^3 - z^2 + 4z - 1}$$
$$= z^3 + 4z^4 + 15z^5 + 59z^6 + 233z^7 + 918z^8$$
$$+ 3616z^9 + 14245z^{10} + 56118z^{11} + 221075z^{12}$$
$$+ 870917z^{13} + 3430947z^{14}$$
$$+ 13516096z^{15} + \cdots$$

The regular expression and generating function for not finding the pattern anywhere is
$$R = (0+2+3+11^*(0+2+3(0+2+3)))^*(11^*3?)?$$

$$G = -\frac{z^2 + 1}{3z^3 - z^2 + 4z - 1}$$
$$= 1 + 4z + 16z^2 + 63z^3 + 248z^4 + 977z^5 + 3849z^6$$
$$+ 15163z^7 + 59734z^8 + 235320z^9 + 927035z^{10}$$
$$+ 3652022z^{11} + 14387013z^{12} + 56677135z^{13}$$
$$+ 223277593z^{14} + 879594276z^{15} + \cdots$$

Problem 62. Construct an automaton that accepts quaternary words where the pattern 131 occurs at the end of the word and anywhere before the end. Find the corresponding regular expression and generating function. Also find the regular expression and generating function for quaternary words that do not contain the pattern at the end.

Answer.

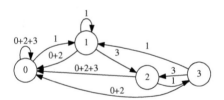

$$R = (0+2+3+1(1+3(13)^*11)^*(0+2+3(13)^*(0+2+3+1(0+2))))^*1(1+3(13)^*11)^*3(13)^*1$$

$$G = -\frac{z^3}{4z - 1}$$
$$= z^3 + 4z^4 + 16z^5 + 64z^6 + 256z^7 + 1024z^8$$
$$+ 4096z^9 + 16384z^{10} + 65536z^{11}$$
$$+ 262144z^{12} + 1048576z^{13}$$
$$+ 4194304z^{14} + 16777216z^{15} + \cdots$$

The regular expression and generating function for not finding the pattern at the end is
$R = (0+2+3+1(1+3(13)*11)*(0+2+3(13)*(0+2+3+1(0+2))))*(1(1+3(13)*11)*(3(13)*)?)?$

$$G = \frac{z^3 - 1}{4z - 1}$$
$$= 1 + 4z + 16z^2 + 63z^3 + 252z^4 + 1008z^5$$
$$+ 4032z^6 + 16128z^7 + 64512z^8 + 258048z^9$$
$$+ 1032192z^{10} + 4128768z^{11} + 16515072z^{12}$$
$$+ 66060288z^{13} + 264241152z^{14}$$
$$+ 1056964608z^{15} + \cdots$$

Problem 63. Construct an automaton that accepts quaternary words where the pattern 0123 occurs only at the end of the word. Find the corresponding regular expression and generating function. Also find the regular expression and generating function for quaternary words that do not contain the pattern anywhere.

Answer.

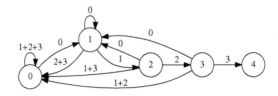

$$R = (1 + 2 + 3 + 0(0 + 1(0 + 20))^*(2 + 3 + 1(1 + 3 + 2(1 + 2))))^*0(0 + 1(0 + 20))^*123$$

$$G = \frac{z^4}{z^4 - 4z + 1}$$
$$= z^4 + 4z^5 + 16z^6 + 64z^7 + 255z^8 + 1016z^9$$
$$+ 4048z^{10} + 16128z^{11} + 64257z^{12}$$
$$+ 256012z^{13} + 1020000z^{14}$$
$$+ 4063872z^{15} + \cdots$$

The regular expression and generating function
for not finding the pattern anywhere is
$$R = (1 + 2 + 3 + 0(0 + 1(0 + 20))^*(2 + 3 + 1(1 + 3 + 2(1 + 2))))^*(0(0 + 1(0 + 20))^*(12?)?)?$$

$$G = \frac{1}{z^4 - 4z + 1}$$
$$= 1 + 4z + 16z^2 + 64z^3 + 255z^4 + 1016z^5$$
$$+ 4048z^6 + 16128z^7 + 64257z^8 + 256012z^9$$
$$+ 1020000z^{10} + 4063872z^{11} + 16191231z^{12}$$
$$+ 64508912z^{13} + 257015648z^{14}$$
$$+ 1023998720z^{15} + \cdots$$

Problem 64. Construct an automaton that accepts quaternary words where the pattern 0123 occurs at the end of the word and anywhere before the end. Find the corresponding regular expression and generating function. Also find the regular expression and generating function for quaternary words that do not contain the pattern at the end.

Answer.

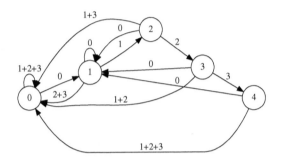

$$R = (1 + 2 + 3 + 0(0 + 1(0 + 2(0 + 30)))^*(2 + 3 + 1(1 + 3 + 2(1 + 2 + 3(1 + 2 + 3)))))^*0(0 + 1(0 + 2(0 + 30)))^*123$$

$$G = -\frac{z^4}{4z - 1}$$
$$= z^4 + 4z^5 + 16z^6 + 64z^7 + 256z^8 + 1024z^9$$
$$+ 4096z^{10} + 16384z^{11} + 65536z^{12}$$
$$+ 262144z^{13} + 1048576z^{14} + 4194304z^{15} + \cdots$$

The regular expression and generating function for not finding the pattern at the end is

$$R = (1 + 2 + 3 + 0(0 + 1(0 + 2(0 + 30)))^*(2 + 3 + 1(1 + 3 + 2(1 + 2 + 3(1 + 2 + 3)))))^*(0(0 + 1(0 + 2(0 + 30)))^*(12?)?)?$$

$$G = \frac{z^4 - 1}{4z - 1}$$
$$= 1 + 4z + 16z^2 + 64z^3 + 255z^4 + 1020z^5$$
$$+ 4080z^6 + 16320z^7 + 65280z^8 + 261120z^9$$
$$+ 1044480z^{10} + 4177920z^{11} + 16711680z^{12}$$
$$+ 66846720z^{13} + 267386880z^{14}$$
$$+ 1069547520z^{15} + \cdots$$

Circular String Patterns

A linear string becomes circular when you attach the last character to the first. The following problems involve automata for recognizing circular strings that contain or do not contain a given pattern. Note that for the purpose of constructing an automaton, a circular string will have an arbitrarily designated first character and direction in which the pattern is read.

Problem 65. Find the automaton and regular expression for circular strings over the alphabet $\{0, 1\}$ that do not contain the pattern 00.

Answer. There are two cases depending on the value of the first character. If the first character is a 1 then the remainder of the string must not contain the 00 pattern. If S is the start state and E

the end or accepting state then this part of the automaton is

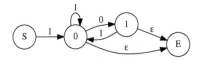

The regular expression is $R = 1(1 + 01)^*(\epsilon + 0)$
In the second case, if the first character is a 0 followed by a 1 then the remainder of the string must not contain the 00 pattern and it must not end in a zero. This part of the automaton is

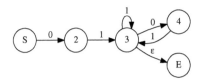

The regular expression is $R = 01(1 + 01)^*$
The combined automaton is

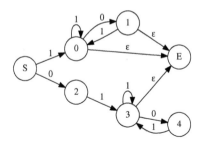

The regular expression is $R = 1(1+01)^*(\epsilon+0) + 01(1+01)^*$

Problem 66. Find the automaton and regular expression for circular strings over the alphabet $\{0,1\}$ that do not contain the pattern 000.

Answer. If the first character is 1 then the rest of the string must not contain 000. The automaton and regular expression for this is

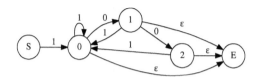

$R = 1(1 + 01 + 001)^*(\epsilon + 0 + 00)$
If the first character is 0 followed by a 1 then
the rest of the string must not contain 000 and
must not end with two 0's. The automaton and
regular expression for this is

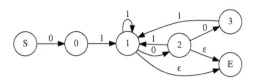

$R = 01(1 + 01 + 001)^*(\epsilon + 0)$
If the first 3 characters are 001 then the rest of
the string must not contain 000 and must not end
with a 0. The automaton and regular expression
for this is

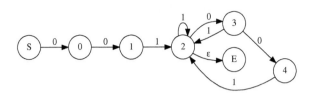

$R = 001(1 + 01 + 001)^*$
To get the complete automaton, combine all the

S states and all the E states in the above automata. The regular expression is the union of the 3 regular expressions:
$$R = 1(1 + 01 + 001)^*(\epsilon + 0 + 00) + 01(1 + 01 + 001)^*(\epsilon + 0) + 001(1 + 01 + 001)^*$$

Problem 67. Find the automaton and regular expression for circular strings over the alphabet $\{0, 1\}$ that do not contain the pattern 011.

Answer. If the first character is 0 then the rest of the string must not contain 11. The automaton and regular expression for this is

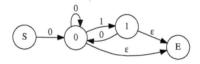

$R = 0(0 + 10)^*(\epsilon + 1)$

If the first two characters are 10 then the rest of the string must not contain 11 and must not end in 1. The automaton and regular expression for this is

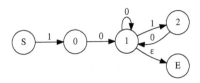

$R = 10(0 + 10)^*$

If the first 2 characters are 11 then the string must have all 1's. The automaton and regular expression for this is

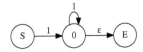

$R = 11^*$

To get the complete automaton, combine all the S states and all the E states in the above 3 automata. The regular expression is the union of the 3 regular expressions:

$R = 0(0 + 10)^*(\epsilon + 1) + 10(0 + 10)^* + 11^*$

Problem 68. Find the automaton and regular expression for circular strings over the alphabet $\{0, 1\}$ that contain the pattern 00.

Answer. Finding circular strings that contain the pattern 00 will have the following automata in parallel

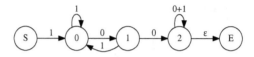

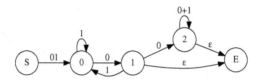

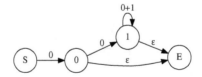

whose regular expressions are respectively
$$R = 1(1 + 01)^*00(0 + 1)^*$$
$$R = 01(1 + 01)^*(0 + 00(0 + 1)^*)$$

$R = 00(0 + 1)^* + 0$

To get the complete automaton, combine all the S states and all the E states in the above 3 automata. The regular expression is the union of the 3 regular expressions:

$R = 1(1+01)^*00(0+1)^* + 01(1+01)^*(0+00(0+1)^*) + 00(0+1)^* + 0$

Problem 69. Find the automaton and regular expression for circular strings over the alphabet $\{0, 1\}$ that contain the pattern 000.

Answer. Finding circular strings that contain the pattern 000 will have 4 parallel automata

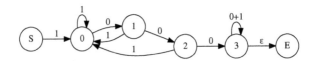

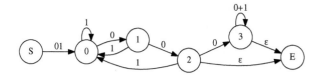

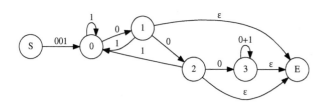

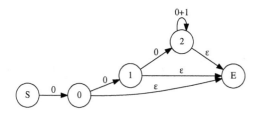

whose regular expressions are respectively
$R = 1(1 + 01 + 001)^*000(0 + 1)^*$
$R = 01(1 + 01 + 001)^*(00 + 000(0 + 1)^*)$

$R = 001(1 + 01 + 001)^*(0 + 00 + 000(0 + 1)^*)$
$R = 000(0 + 1)^* + 00 + 0$

To get the complete automaton, combine all the S states and all the E states in the above 4 automata. The regular expression is the union of the 4 regular expressions:

$R = 1(1 + 01 + 001)^*000(0 + 1)^* + 01(1 + 01 + 001)^*(00 + 000(0 + 1)^*) + 001(1 + 01 + 001)^*(0 + 00 + 000(0 + 1)^*) + 000(0 + 1)^* + 00 + 0$

Miscellaneous Problems

Problem 70. A string is constructed from the digits 1 through 6 in such a way that the first number is always 1 and each subsequent number must be less than or equal to one greater than the largest number that has appeared so far. The string terminates when the number 6 appears. Create an automaton and regular expression for recognizing such strings. Give an example of a real process that generates such strings.

Answer. The automaton for this problem is

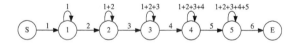

and its regular expression is

$R = 11^*2(1+2)^*3(1+2+3)^*4(1+2+3+4)^*5(1+2+3+4+5)^*6$

An example of a real process that would generate a string like this is rolling a die until all six faces turn up at least once. The numbers 1 through 6 represent how many unique faces have been rolled so far. They do not represent the numbers on the die faces. The first roll produces the first unique face, so the first character is always a 1. The second roll may be the same as the first in which case the next string character is again a 1, or it may be a new face in which case the next character is a 2. On each roll, you can get at most one more unique face, so the next character can be at most one more than the largest so far. Another example is collecting baseball cards in cereal boxes.

Problem 71. Find the generating function for the number of strings of a given length in the previous problem.

Answer.

$$G = \frac{z^6}{(1-z)(1-2z)(1-3z)(1-4z)(1-5z)}$$
$$= z^6 + 15z^7 + 140z^8 + 1050z^9 + 6951z^{10}$$
$$+ 42525z^{11} + 246730z^{12} + 1379400z^{13}$$
$$+ 7508501z^{14} + 40075035z^{15} + \cdots$$

Note that the coefficients of this generating function expansion $(1, 15, 140, 1050, 6951, \ldots)$ are the Stirling numbers of the second kind.

Problem 72. Spike and Spud each have 3 dollars and decide to play a coin tossing game. Each time they toss a coin, if the result is heads (H) then Spud gives Spike one dollar. If the result is tails (T) then Spike gives Spud one dollar. The game continues until one of them has no money. Model this game as an automaton where the input is the result of the coin tosses and the states are the number of dollars Spike has.

Answer. The automaton is

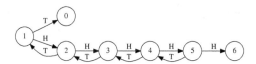

Spike starts the game in state 3 and the game ends either in state 0 (Spike has lost all his money) or in state 6 (Spud has lost all his money).

Problem 73. In the previous problem, find the regular expressions for the case where Spike ends in state 0 and where he ends in state 6.

Answer. Start by eliminating states 1 and 5

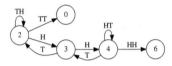

Next eliminate states 2 and 4

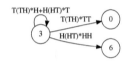

Now you can read off the regular expressions. The regular expression for starting in state 3 and ending in state 0 is

$R = (T(TH)^*H + H(HT)^*T)^*T(TH)^*TT$

The regular expression for starting in state 3 and ending in state 6 is

$R = (T(TH)^*H + H(HT)^*T)^*H(HT)^*HH$

These regular expressions reveal some of the dynamics of the game. Both of them start with the term $(T(TH)^*H + H(HT)^*T)^*$ which represents returns to the start, state 3. At some point the game is in state 3 for the last time before it ends in state 0 or 6. If it finally leaves state 3 on a T then it goes to state 2 where it may oscillate for a time between states 2 and 1. This part of the dynamics is represented by the term $T(TH)^*$. Finally there is a TT which ends the game in state 0. So the regular expression for moving away from state 3 for the last time and ending in state 0 is $T(TH)^*TT$.

Problem 74. For the previous problem, find how many games of a particular length are possible by deriving the generating function from the regular expressions.

Answer. The regular expressions for losing (ending in state 0) and winning (ending in state 6) are symmetric under an exchange of the symbols H and T so they will both have the same number of strings of a given length. Using the usual procedure for converting a regular expression into a generating function gives

$$G = \frac{z^3}{1 - 3z^2}$$
$$= z^3 + 3z^5 + 9z^7 + 27z^9 + 81z^{11}$$
$$+ 243z^{13} + 729z^{15} + \cdots$$

So the number of games of length $2n + 1$ is 3^{n-1}. Do you see why it is not possible to have a game of even length?

Problem 75. Create an automaton for a string of correctly nested parentheses if the nesting level never goes beyond 5. The string is correctly nested if the number of right (closing) parentheses never exceeds the number of left (opening) parentheses. Use the symbol r for a right parenthesis and the symbol l for a left parenthesis.

Answer. The automaton is

The start and end state is state 0. The automaton is in state n when n more l's than r's have been read.

Problem 76. Find the regular expression for the automaton in the previous problem, i.e. find a regular expression for recognizing strings of correctly nested parentheses if the nesting can go no more than 5 deep.

Answer. The regular expression is
$$R = (l(l(l(l(l(lr)^*r)^*r)^*r)^*r)^*$$
The following is a list of the length 10 strings that are recognized.

```
((((()))))     (((()())))     (((())()))
(((()))())     (((())))()     ((()(())))
((()()()))     ((()())())     ((()()))()
((())(()))     ((())()())     ((())())()
((()))(())     ((()))()()     (()((())))
(()(()()))     (()(())())     (()(()))()
(()()(()))     (()()()())     (())((()))
(()())(())     (())(())()     (())()(())
(())(()())     ()(((())))     ()((()()))
(())()()()     ()((()))()     ()(()(()))
()((())())     ()(()())()     ()(())(())
()(()()())     ()()((()))     ()()(()())
()(())()()     ()()(())()     ()()()(())
()()(())()     ()()()(())     ()()()()()
```

Problem 77. For the previous problem find how many strings of a given length are possible by deriving the generating function from the regular expression.

Answer. Using the usual procedure for converting a regular expression into a generating function produces

$$G(z) = \cfrac{1}{1 - \cfrac{z^2}{1 - \cfrac{z^2}{1 - \cfrac{z^2}{1 - \cfrac{z^2}{1 - z^2}}}}}$$

which simplifies to

$$G(z) = -\frac{3z^4 - 4z^2 + 1}{z^6 - 6z^4 + 5z^2 - 1}$$
$$= 1 + z^2 + 2z^4 + 5z^6 + 14z^8 + 42z^{10}$$
$$+ 131z^{12} + 417z^{14} + \cdots$$

Problem 78. If we remove the limit of 5 for the nesting depth in the previous problem, then the automaton must have an infinite number of states. The regular expression would also be infinitely long, so it can't be written out but it can be defined recursively. In other words it can be defined in terms of itself as a recursion equation. Find the recursion equation and use it to find the generating function for the number of strings of a given length that the infinite automaton will accept.

Answer. The infinite automaton would look like the one above except that it would extend to an infinite number of states instead of stopping at 5.

Since the automaton is infinite, if you remove state 0 what you have left still looks like the original automaton. Adding the state back involves adding an l to the beginning of the string, an r to the end of the string, and starring the result. The regular expression can therefore be defined recursively as

$$R = (lRr)^*$$

Let $G(z)$ be the result of turning R into a generating function then the recursive equation for R gets transformed into a recursive equation for $G(z)$:

$$G(z) = \frac{1}{1 - z^2 G(z)}$$

which can be solved to give

$$\begin{aligned} G(z) &= \frac{1 - \sqrt{1 - 4z^2}}{2z^2} \\ &= 1 + z^2 + 2z^4 + 5z^6 + 14z^8 + 42z^{10} \\ &\quad + 132z^{12} + 429z^{14} + 1430z^{16} \\ &\quad + 4862z^{18} + 16796z^{20} + \cdots \end{aligned}$$

The coefficients in the expansion are the well known Catalan numbers.

Problem 79. Spike enters a casino with no money but he convinces them to give him an unlimited credit line. He plays a game that is equivalent to a coin toss where he wins a dollar on heads and loses a dollar on tails. Create an automaton where the inputs are the symbols h and t, corresponding to the results of the coin toss, and the states are the amount of money Spike has. What is the regular expression for the event that he leaves the casino with just as much money as when he arrived, i.e. zero dollars.

Answer. The automaton is similar to the one in the problem on page 123 where Spike plays a coin tossing game with Spud and they both start with 3 dollars. The difference is that Spike starts in state 0 and there is no limit to how much money he can lose or win. The automaton must have an infinite number of negative and positive states. A small portion of the automaton is shown below, corresponding to the first two positive and negative states.

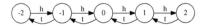

The regular expression for an infinite state automaton would also be infinite, so it can't be written out but it can be defined recursively as follows:

$$R = (hAt + tBh)^*$$
$$A = (hAt)^*$$
$$B = (tBh)^*$$

In the expression for R the term hAt represents an excursion of any length among the positive states that eventually returns to 0 while the term tBh represents the same thing for the negative states. The term A is the regular expression for a return to state 1 after a visit to more positive states. The term B is the regular expression for a return to state -1 after a visit to more negative states. Note the similarity of these terms to the regular expression for an unlimited number of nested parentheses defined in a previous problem. Also note that a recursive regular expression can usually be easily converted into what is called a context free grammar (see one of the references for a full description of context free grammars).

In this case the context free grammar is:

$$S \to hAtS$$
$$S \to tBhS$$
$$S \to \epsilon$$
$$A \to hAtA$$
$$A \to \epsilon$$
$$B \to tBhB$$
$$B \to \epsilon$$

These are called production rules. The start symbol is S which can be replaced using any of its production rules. The symbols A and B can likewise be replaced using any of their production rules. The process continues until only the h and t symbols remain. In this way all the strings accepted by the recursive regular expression can be generated.

Problem 80. In the previous problem find the generating function for the number of ways Spike can play n games and leave with no money.

Answer. The generating function is found by converting the regular expression. This gives:

$$G(z) = \frac{1}{1 - z^2 A(z) - z^2 B(z)}$$

$$A(z) = \frac{1}{1 - z^2 A(z)}$$

$$B(z) = \frac{1}{1 - z^2 B(z)}$$

Solving for $A(z)$ and $B(z)$ and substituting into the expression for $G(z)$ produces

$$\begin{aligned} G(z) &= \frac{1}{\sqrt{1 - 4z^2}} \\ &= 1 + 2z^2 + 6z^4 + 20z^6 + 70z^8 + 252z^{10} \\ &\quad + 924z^{12} + 3432z^{14} + 12870z^{16} \\ &\quad + 48620z^{18} + \cdots \end{aligned}$$

The coefficients in the expansion are the central binomial coefficients given by the formula $\binom{2n}{n}$. The generating function says there are 70 ways that Spike can play 8 games and leave with no money.

Problem 81. Continuing from the previous two problems, suppose that Spike ends up leaving with one dollar. Find a recursive regular expression for this event and the generating function that counts the number of ways it can happen.

Answer. At some point Spike will be at zero for the last time as he wins a dollar and then stays positive before finally ending at one dollar. The recursive regular expression for this is

$$R = (hAt + tBh)^* hA$$
$$A = (hAt)^*$$
$$B = (tBh)^*$$

This is the same as the previous expression except for the hA term that indicates a transition to state 1 followed by a possible excursion to higher states that ends back in state 1. The generating function is:

$$G(z) = \frac{1 - \sqrt{1 - 4z^2}}{2z\sqrt{1 - 4z^2}}$$
$$= z + 3z^3 + 10z^5 + 35z^7 + 126z^9 + 462z^{11}$$
$$+ 1716z^{13} + 6435z^{15} + 24310z^{17}$$
$$+ 92378z^{19} + \cdots$$

The formula for the coefficient of z^{2n+1} in the expansion is $\binom{2n+1}{n}$. The generating function shows that there are 126 ways that Spike can play 9 games and come out one dollar ahead. The corresponding context free grammar for this problem is

$$S \to hAtS$$
$$S \to tBhS$$
$$S \to hA$$
$$A \to hAtA$$
$$A \to \epsilon$$
$$B \to tBhB$$
$$B \to \epsilon$$

Problem 82. Continuing from the previous problem, suppose Spike ends up leaving with two dollars. Find a recursive regular expression for this event and the generating function that counts the number of ways it can happen.

Answer. Analogous to the previous problem, it is clear that the recursive regular expression must be:

$$R = (hAt + tBh)^*hAhA$$
$$A = (hAt)^*$$
$$B = (tBh)^*$$

This is the same as the previous problem except for an additional hA term indicating a transition to state 2 followed by a possible excursion to higher states, ending in state 2. The generating function is:

$$G(z) = \left(\frac{1 - \sqrt{1 - 4z^2}}{2z}\right)^2 \frac{1}{\sqrt{1 - 4z^2}}$$
$$= z^2 + 4z^4 + 15z^6 + 56z^8 + 210z^{10} + 792z^{12}$$
$$+ 3003z^{14} + 11440z^{16} + 43758z^{18}$$
$$+ 167960z^{20} + \cdots$$

The formula for the coefficient of z^{2n+2} in the expansion is $\binom{2n+2}{n}$. The generating function shows that there are 210 ways that Spike can play 10 games and come out two dollars ahead.

Problem 83. For the previous problem, find a recursive regular expression and generating function for the event that Spike leaves with zero or more dollars.

Answer. Now there can be zero or more hA terms at the end so the recursive regular expression must be:

$$R = (hAt + tBh)^*(hA)^*$$
$$A = (hAt)^*$$
$$B = (tBh)^*$$

and the generating function is:

$$G(z) = \frac{2z}{2z - 1 + \sqrt{1 - 4z^2}} \frac{1}{\sqrt{1 - 4z^2}}$$
$$= 1 + z + 3z^2 + 4z^3 + 11z^4 + 16z^5 + 42z^6$$
$$+ 64z^7 + 163z^8 + 256z^9 + 638z^{10}$$
$$+ 1024z^{11} + 2510z^{12} + 4096z^{13}$$
$$+ 9908z^{14} + 16384z^{15} + 39203z^{16}$$
$$+ 65536z^{17} + 155382z^{18} + 262144z^{19}$$
$$+ 616666z^{20} + \cdots$$

The generating function shows that there are 638 ways that Spike can play 10 games and come out with zero or more dollars.

Problem 84. Find a recursive regular expression for matching regular expressions over the binary alphabet $\{a, b\}$. Assume the regular expression uses $\cup$ instead of $+$, $\#$ instead of $*$, and square brackets instead of parentheses. The recursive regular expression should match all syntactically correct regular expressions except for the empty string ϵ.

Answer. A regular expression can be broken down into a union of smaller terms. Let T be a term, then the regular expression will be $R = (T\cup)^*T$. Each T can be a product of any number of factors. Let F represent a factor, then $T = F^*F$.

Each factor will be a starred or an unstarred expression. Let X be an expression, then $F = X + X^{\#}$. An expression will be one of the alphabet letters or a parenthesized regular expression, $X = a+b+[R]$. If we eliminate all of the variables except for F then the regular expression can be defined as

$$R = (F^*F\cup)^*F^*F$$
$$F = a + a^{\#} + b + b^{\#} + [R] + [R]^{\#}$$

This regular expression corresponds to the following context free grammar

$$R \rightarrow T \cup R$$
$$R \rightarrow T$$
$$T \rightarrow FT$$
$$T \rightarrow F$$
$$F \rightarrow X^{\#}$$
$$F \rightarrow X$$
$$X \rightarrow [R]$$
$$X \rightarrow a$$
$$X \rightarrow b$$

Now let's find the generating function for the number of syntactically correct regular expressions of a given length. Let $R(z)$ be the generat-

ing function, and let $F(z)$ be the algebraic form
of F. Then

$$R(z) = \frac{1}{1 - \frac{zF}{1-F}}$$
$$= \frac{F(z)}{1 - (1+z)F(z)}$$

and

$$F(z) = z^2 R(z) + z^3 R(z) + 2z + 2z^2$$
$$= z^2(1+z)R(z) + 2z(1+z)$$

Substituting $F(z)$ into the expression for $R(z)$
and solving for $R(z)$ gives

$$R(z) =$$
$$\frac{1 - 2z - 5z^2 - 3z^3 - \sqrt{z^6 + 6z^5 + 13z^4 + 6z^3 - 6z^2 - 4z + 1}}{2z^2(1+z)^2}$$

The Taylor series expansion is

$$R(z) = 2z + 6z^2 + 22z^3 + 84z^4 + 328z^5 + 1318z^6$$
$$+ 5400z^7 + \cdots$$

The 2 regular expressions of length 1 are

a b

The 6 regular expressions of length 2 are

ab ba aa bb a^* b^*

The 22 regular expressions of length 3 are

(a) (b) aa^* ba^* ab^* bb^* $a+a$ $b+a$

a^*a b^*a aaa baa aba bba $a+b$ $b+b$

a^*b b^*b aab bab abb bbb

The 84 regular expressions of length 4 are

(a^*)	(b^*)	$a(a)$	$b(a)$	(aa)	(ba)
$a(b)$	$b(b)$	(ab)	(bb)	$(a)^*$	$(b)^*$
$a+a^*$	$b+a^*$	a^*a^*	b^*a^*	aaa^*	baa^*
aba^*	bba^*	$a+b^*$	$b+b^*$	a^*b^*	b^*b^*
aab^*	bab^*	abb^*	bbb^*	$(a)a$	$(b)a$
a^*+a	b^*+a	$aa+a$	$ba+a$	$ab+a$	$bb+a$
aa^*a	ba^*a	ab^*a	bb^*a	$a+aa$	$b+aa$
a^*aa	b^*aa	$aaaa$	$baaa$	$abaa$	$bbaa$
$a+ba$	$b+ba$	a^*ba	b^*ba	$aaba$	$baba$
$abba$	$bbba$	$(a)b$	$(b)b$	a^*+b	b^*+b
$aa+b$	$ba+b$	$ab+b$	$bb+b$	aa^*b	ba^*b
ab^*b	bb^*b	$a+ab$	$b+ab$	a^*ab	b^*ab
$aaab$	$baab$	$abab$	$bbab$	$a+bb$	$b+bb$
a^*bb	b^*bb	$aabb$	$babb$	$abbb$	$bbbb$

Keep in mind that these regular expressions are all syntactically correct, but don't all necessarily make sense.

Appendix A: Grep Format Conversion

Grep[1] is a useful program for trying out the regular expressions in this book. The regular expressions in this book are written in the customary notation of computer science. To convert this notation to something grep can understand, do the following:

- Convert $+$ to $|$

- Convert $(\epsilon + x)$ to $x?$

and of course the *'s are not superscripted. So for example
$$1(1+01+001)^*(\epsilon+0+00)+01(1+01+001)^*(\epsilon+0)+$$

---

[1] grep is part of Unix related operating systems, and is available for Redmond, Washington originating ones as part of the Cygwin package (among others).

$001(1 + 01 + 001)^*$
becomes
$1(1|01|001) * (0|00)?|01(1|01|001) * 0?|001(1|01|001)*$

If you want to match only whole lines with a given regular expression, remember to insert ˆ at the start of the regular expression and $ at the end. For example if you have a file containing all possible binary 10 bit strings (one per line), and you want to find all those strings (not substrings) that match the regular expression
$(0(0(0(0(01) * 1) * 1) * 1) * 1)*$
then the command (in Linux at least) is:

```
egrep '^(0(0(0(0(01)*1)*1)*1)*$' b2w10.dat
```

where `egrep`, as opposed to `grep` is used for extended regular expression syntax. In this example, the only reason you need `egrep` is so you don't have to escape the parentheses with a backslash, although it's a good idea to always use `egrep` for the regular expressions in this book.

Further Reading

- *Regular Algebra and Finite Machines*, John Horton Conway, 2012

- *Implementing Regular Expressions* webpage, Russ Cox

- *Mastering Regular Expressions*, Jeffrey E. F. Friedl, 3rd ed, 2006

- *Introduction to Automata Theory, Languages, and Computation*, Hopcroft and Ullman, 1979

- *Automata, Computability and Complexity : Theory and Applications*, Elaine Rich, 2008

- *Introduction to the Theory of Computation*, Michael Sipser, 3rd ed, 2013

Acknowledgements

In ordinary life we hardly realize that we receive a great deal more than we give, and that it is only with gratitude that life becomes rich. It is very easy to overestimate the importance of our own achievements in comparison with what we owe to others.

Dietrich Bonhoeffer, letter to parents from prison, Sept. 13, 1943

We'd like to thank our parents for helping us in many ways.

We thank the makers and maintainers of all the software we've used in the production of this book, including: the Emacs text editor, the LaTex typesetting system, LaTeXML, Inkscape, Evince document viewer, POV-Ray, Maxima computer algebra system, gcc, Guile, awk, sed, bash shell, and the Linux operating system. Particular thanks to Friedrich A. Lohmueller for his POV-Ray tutorials.

About the Authors

Stefan Hollos and **J. Richard Hollos** are physicists by training, and enjoy anything related to math, physics, and computing. They are the authors of

- **Probability Problems and Solutions**

- **Combinatorics Problems and Solutions**

- **The Coin Toss: The Hydrogen Atom of Probability**

- **Pairs Trading: A Bayesian Example**

- **Simple Trading Strategies That Work**

- **Bet Smart: The Kelly System for Gambling and Investing**

- **The QuantWolf Guide to Calculating Bond Default Probabilities**

- The Mathematics of Lotteries: How to Calculate the Odds

- Signals from the Subatomic World: How to Build a Proton Precession Magnetometer

They are brothers and business partners at Exstrom Laboratories LLC in Longmont, Colorado. Their website is exstrom.com

Thank You

Thank you for buying this book.

Sign up for the Abrazol Publishing Newsletter and receive news on updates, new books, and special offers. Just go to

http://www.abrazol.com/

and enter your email address.

www.ingramcontent.com/pod-product-compliance
Lightning Source LLC
LaVergne TN
LVHW022321060326
832902LV00020B/3598